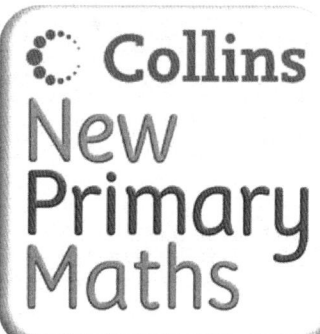

Collins New Primary Maths

Assessment Pack 5

Author: Peter Clarke

William Collins' dream of knowledge for all began with the publication of his first book in 1819. A self-educated mill worker, he not only enriched millions of lives, but also founded a flourishing publishing house. Today, staying true to this spirit, Collins books are packed with inspiration, innovation and practical expertise. They place you at the centre of a world of possibility and give you exactly what you need to explore it.

Collins. Freedom to Teach.

Published by Collins
An imprint of HarperCollinsPublishers
77 – 85 Fulham Palace Road
Hammersmith
London
W6 8JB

Browse the complete Collins catalogue at
www.collinseducation.com

© HarperCollinsPublishers Limited 2007

10 9

ISBN-13 978 0 00 722052 6

Peter Clarke asserts his moral right to be identified as the author of this work

British Library Cataloguing in Publication Data
A Catalogue record for this publication is available from the British Library

Cover design by Laing&Carroll
Cover artwork by Jonatronix Ltd
Internal design and page make-up by Neil Adams
Illustrations by Neil Adams and Bridget Dowty
Edited by Ros Davies and Chris Davies

Printed and bound by Martins the Printers, Berwick-upon-Tweed

Mixed Sources
Product group from well-managed
forests and other controlled sources
www.fsc.org Cert no. SW-COC-1806
© 1996 Forest Stewardship Council
FSC

FSC is a non-profit international organisation established to promote the responsible management of the world's forests. Products carrying the FSC label are independently certified to assure consumers that they come from forests that are managed to meet the social, economic and ecological needs of present and future generations.

Find out more about HarperCollins and the environment at
www.harpercollins.co.uk/green

Contents

Introduction

What does the Primary National Strategy (PNS) *Renewed Framework for Mathematics* (2006) say about assessment?

The PNS *Renewed Framework for Mathematics* identifies two main purposes of assessment:

- Assessment *for* learning (formative on-going assessment)
- Assessment *of* learning (summative assessment)

Assessment *for* learning involves both pupils and teachers finding out about the specific strengths and weaknesses of individual children, and the class as a whole, and using this to inform future teaching and learning.

Assessment *for* learning:

- is part of the planning process
- is informed by learning objectives
- engages children in the assessment process
- recognises the achievements of all children
- takes account of how children learn
- motivates learners.

Assessment *of* learning is any assessment that summarises where individual children, and the class as a whole, are at a given point in time. It provides a snapshot of what has been learned.

The *Collins NEW Primary Maths (CNPM)* Assessment Packs

The *CNPM Assessment Packs* aim to provide guidance in both Assessment *for* learning and Assessment *of* learning.

The *CNPM Assessment Packs* consist of three key features:

- Section 1: Adult Directed Tasks
- Section 2: Pupil Self assessments
- Section 3: Tests

Section 1: Adult Directed Tasks

Purposes

- To assist in identifying particular children's strengths and weaknesses.
- To inform future planning and teaching of individual children and the class as a whole.
- To provide some guidance about what to do for those children who are achieving above or below expectations.

When to use this feature

- Anytime throughout the year when you are uncertain about a child's, or a group of children's, understanding of a particular objective.

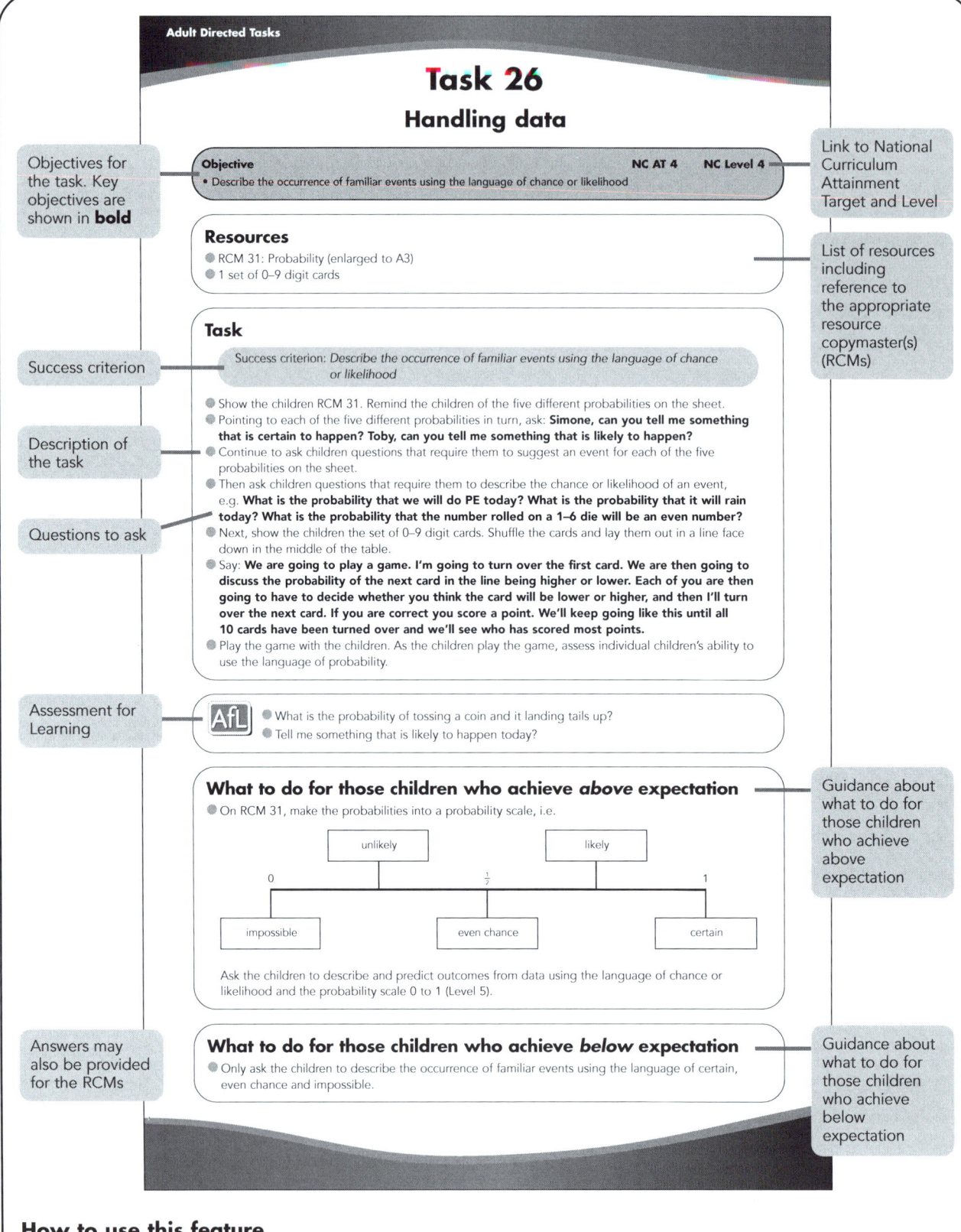

Adult Directed Tasks

Task 26
Handling data

Objectives for the task. Key objectives are shown in **bold**

Objective
NC AT 4   NC Level 4
• Describe the occurrence of familiar events using the language of chance or likelihood

Link to National Curriculum Attainment Target and Level

Resources
● RCM 31: Probability (enlarged to A3)
● 1 set of 0–9 digit cards

List of resources including reference to the appropriate resource copymaster(s) (RCMs)

Task

Success criterion

Success criterion: *Describe the occurrence of familiar events using the language of chance or likelihood*

Description of the task

Questions to ask

● Show the children RCM 31. Remind the children of the five different probabilities on the sheet.
● Pointing to each of the five different probabilities in turn, ask: **Simone, can you tell me something that is certain to happen? Toby, can you tell me something that is likely to happen?**
● Continue to ask children questions that require them to suggest an event for each of the five probabilities on the sheet.
● Then ask children questions that require them to describe the chance or likelihood of an event, e.g. **What is the probability that we will do PE today? What is the probability that it will rain today? What is the probability that the number rolled on a 1–6 die will be an even number?**
● Next, show the children the set of 0–9 digit cards. Shuffle the cards and lay them out in a line face down in the middle of the table.
● Say: **We are going to play a game. I'm going to turn over the first card. We are then going to discuss the probability of the next card in the line being higher or lower. Each of you are then going to have to decide whether you think the card will be lower or higher, and then I'll turn over the next card. If you are correct you score a point. We'll keep going like this until all 10 cards have been turned over and we'll see who has scored most points.**
● Play the game with the children. As the children play the game, assess individual children's ability to use the language of probability.

Assessment for Learning

AfL
● What is the probability of tossing a coin and it landing tails up?
● Tell me something that is likely to happen today?

What to do for those children who achieve *above* expectation
● On RCM 31, make the probabilities into a probability scale, i.e.

unlikely		likely	

0	$\frac{1}{2}$	1

impossible	even chance	certain

Ask the children to describe and predict outcomes from data using the language of chance or likelihood and the probability scale 0 to 1 (Level 5).

Guidance about what to do for those children who achieve above expectation

Answers may also be provided for the RCMs

What to do for those children who achieve *below* expectation
● Only ask the children to describe the occurrence of familiar events using the language of certain, even chance and impossible.

Guidance about what to do for those children who achieve below expectation

How to use this feature
● Using **Record-keeping format 1: Adult Directed Task assessment sheet**:
  – complete the top section
  – copy the Success criteria from the relevant Adult Directed Task
  – write down the names of the children you are going to be working with.

Record-keeping format 1 Adult Directed Task assessment sheet

Objective(s): _Understand percentage as the number of parts in every 100_   Date: _23/01/2008_   Adult: _Ann Scott_

and express tenths and hundredths as percentages   NC Level: _2_   Class: _5 G_

Child's name	Success criteria				Other observations	Objective(s) achieved	Future action
	Understand percentage as number of parts in every 100	Express tenths and hundredths as percentages					
Aileen Jones							
Carol Roberts							
Simon Chandra							

● Use **Record-keeping format 1** to record individual children's performance during the task, commenting upon particular strengths and weaknesses, how competent you feel the children are with this objective(s) and any future action you may consider appropriate.

Section 2: Pupil Self assessments

Purpose

● To provide children with the opportunity to undertake some form of self assessment at the end of a unit.

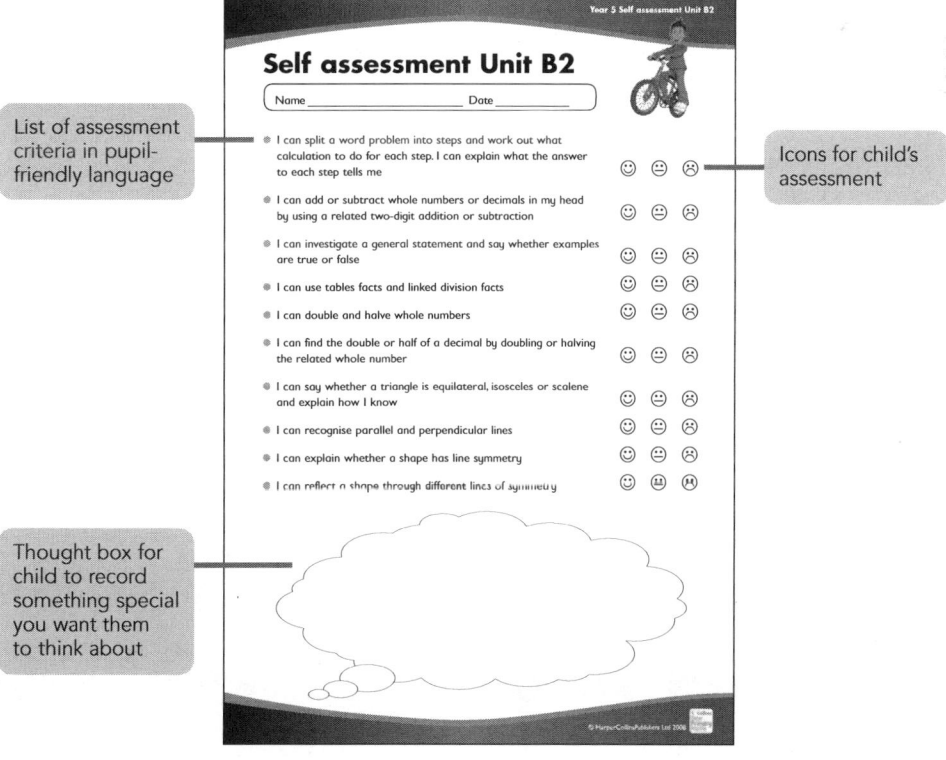

List of assessment criteria in pupil-friendly language

Icons for child's assessment

Thought box for child to record something special you want them to think about

When to use this feature

- At the end of each unit.

How to use this feature

- Distribute the relevant **Pupil Self assessment sheet** at the end of the unit.
- The empty thought box at the bottom of the sheet is designed to be used by the children to record anything special that you might like them to think about, e.g.
  – anything they feel they need more practice on
  – what they think they should or could learn next
  – any special equipment that they used to help them during the unit
  – anything they particularly liked or disliked that they did during the unit.
- Ask the children to complete the sheet independently.
- After the children have completed the sheet, as a class, discuss specific objectives, asking individual children to comment on what they have written.

Section 3: Tests

Purposes

- To provide an indication of how individual children, and the class as a whole, have performed during a term, and to inform future planning.
- To inform teacher assessment when assigning an overall National Curriculum Level in mathematics.
- To inform teacher assessment when assigning a National Curriculum Level for each Attainment Target in mathematics.

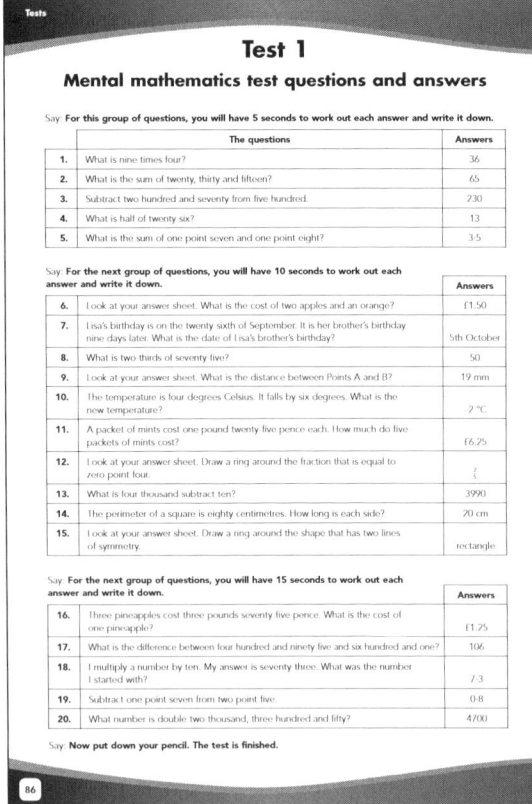

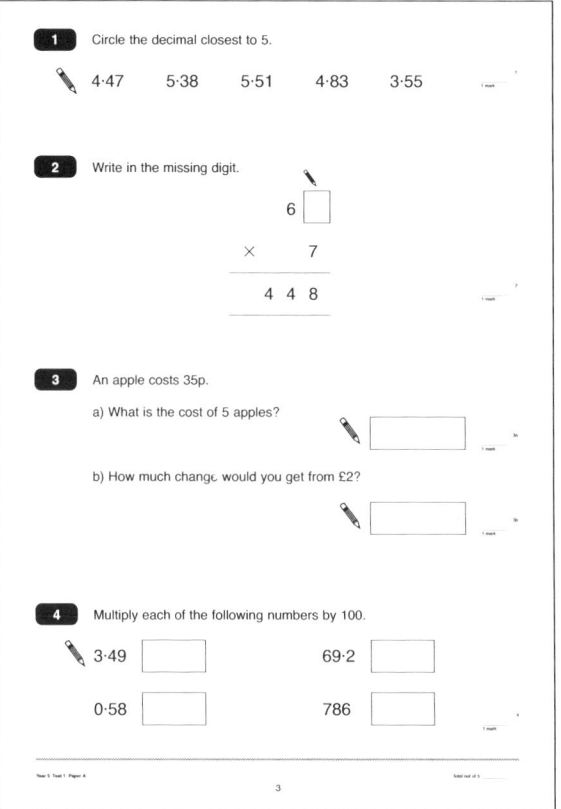

When to use this feature

- At the end of each term.

How to use this feature

- Distribute the relevant Test towards the end of the term. Ensure that all children have the necessary resources.
- Ask the Mental mathematics test questions. This is designed to take approximately 20 minutes.
- Children work independently to complete Papers A and B. Each paper should take approximately 45 minutes.
- Mark the papers and record individual children's results on their paper. You may wish to use **Record-keeping formats 2**, **3** or **4** to analyse the performance of individual children and particular test questions.

Record-keeping format 2 Test 1 grid for test analysis (Mental mathematics test)

Name	Multiplication tables	Addition: more than two numbers	Subtraction	Halving whole numbers	Addition: decimals	Addition: money	Time differences	Fraction of amounts	Distance difference	Temperature difference	Multiplication: money	Fraction and decimal equivalences	Subtraction	Perimeter	2-D shapes/symmetry	Division: money	Difference: near multiples of 100	Multiplying/dividing whole numbers by 10	Subtraction: decimals	Doubling whole numbers	Mental mathematics test score (out of 20)
AT	2	2	2	2	2	1 & 2	1 & 3	2	3	2	1 & 2	2	2	3	3	1 & 2	2	2	2	2	
Question	1	2	3	4	5	6	7	8	9	10	11	12	13	14	15	16	17	18	19	20	
Mark	1	1	1	1	1	1	1	1	1	1	1	1	1	1	1	1	1	1	1	1	
1. Aileen Jones	1	0	1	1	0	0	1	0	1	1	0	1	0	1	0	1	0	1	1	1	13
2. Carol Roberts	1	0	0	1	0	1	0	1	0	0	1	1	0	1	1	1	0	0	0	1	10
3. Simon Chandra	1	1	1	1	1	1	1	1	1	1	1	1	0	1	1	1	1	1	1	0	18
4.																					
5.																					
6.																					
7.																					
8.																					
9.																					
10.																					
11.																					
12.																					
13.																					
14.																					
15.																					
16.																					
17.																					
18.																					
19.																					
20.																					
21.																					
22.																					
23.																					
24.																					
25.																					
26.																					
27.																					
28.																					
29.																					
30.																					
Number correct																					
Number incorrect or omitted																					
Percentage correct																					
Percentage incorrect or omitted																					

- Using the National Curriculum Level Indicator, assign a Level for the Test.

National Curriculum Level Indicator			
Below Level 3	Level 3	Level 4	Level 5
0–22	23–47	48–73	74–90

* These Tests must be seen only as a guide to help gaining an overall best fit in mathematics.

● Use your professional judgement of each child's overall performance during the term in each of the National Curriculum Attainment Targets. Take into account the following:
  – performance in the Test
  – observations made during Adult Directed Tasks
  – mastery of objectives from the PNS *Renewed Framework for Mathematics* (2006)
  – performance in whole class discussions
  – participation in group work
  – work presented in exercise books
  – any other written evidence.

You may wish to use the following record-keeping formats to assign a Level for each of the National Curriculum Attainment Targets:

Record-keeping format	National Curriculum Attainment Target
Record-keeping format 5	Attainment Target 1 – Using and applying mathematics
Record-keeping format 6	Attainment Target 2 – Number and algebra
Record-keeping format 7	Attainment Target 3 – Shape, space and measures
Record-keeping format 8	Attainment Target 4 – Handling data

Once you have decided which Level best fits a particular child you may wish to identify how competent a child is at that Level by using the following key:

C	Becoming competent at this Level (Achieving up to $\frac{1}{3}$ of the Level Descriptors)	Lower
B	Competent at this Level (Achieving between $\frac{1}{3}$ and $\frac{2}{3}$ of the Level Descriptors)	Secure
A	Very competent at this Level (Achieving more than $\frac{2}{3}$ of the Level Descriptors)	Upper

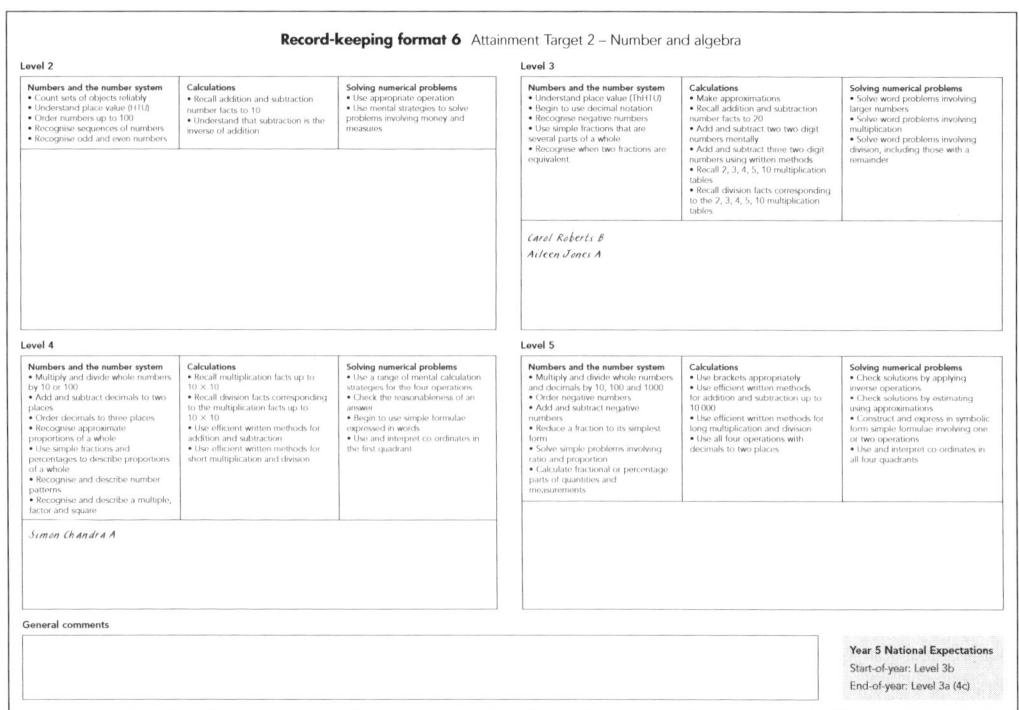

● You may also wish to record children's attainment in each of the Key objectives (also referred to as end-of-year expectations) using either of the following record-keeping formats:

Record-keeping format 9: Class record of the end-of-year expectations

Record-keeping format 9 Class record of the end-of-year expectations

Class: *5G*
Date: *09/05/2008*

**Year 5
End-of-year expectations**

Names

Aileen Jones
Carol Roberts
Simon Chandra

Counting and understanding number (AT2)
Explain what each digit represents in whole numbers and decimals with up to two places, and partition, round and order these numbers (Level 4)

Knowing and using number facts (AT2)
Use knowledge of place value and addition and subtraction of two-digit numbers to derive sums and differences and doubles and halves of decimals, e.g. 6·5 ± 2·7, halve 5·6, double 0·34 (Level 4)

Calculating (AT2)
Use efficient written methods to add and subtract whole numbers and decimals with up to two places (Level 4)

Understanding shape (AT3)
Read and plot co-ordinates in the first quadrant; recognise parallel and perpendicular lines in grids and shapes; use a set-square and ruler to draw shapes with perpendicular or parallel sides (Level 4)

Measuring (AT3)
Draw and measure lines to the nearest millimetre; measure and calculate the perimeter of regular and irregular polygons; use the formula for the area of a rectangle to calculate the rectangle's area (Level 4)

Handling data (AT4)
Construct frequency tables, pictograms and bar and line graphs to represent the frequencies of events and changes over time (Level 4)

NOTES: **Using and applying mathematics (AT1)** is incorporated throughout
End-of-year National Expectations: Level 3a (4c)

Record-keeping format 10: Individual child's record of end-of-year expectations

Record-keeping format 10 Individual child's record of the end-of-year expectations

Name: *Simon Chandra*

Year 4		Year 5		Year 6	Year 6 progression to Year 7
Counting and understanding number (AT2)					
Use diagrams to identify equivalent fractions; interpret mixed numbers and position them on a number line (Level 3)	A	Explain what each digit represents in whole numbers and decimals with up to two places, and partition, round and order these numbers (Level 4)	A	Express one quantity as a percentage of another; find equivalent percentages, decimals and fractions (Level 4)	Use ratio notation, reduce a ratio to its simplest form and divide a quantity into two parts in a given ratio; solve simple problems involving ratio and direct proportion (Level 5)
Knowing and using number facts (AT2)					
Derive and recall multiplication facts up to 10 × 10, the corresponding division facts and multiples of numbers to 10 up to the tenth multiple (Level 4)	B	Use knowledge of place value and addition and subtraction of two-digit numbers to derive sums and differences and doubles and halves of decimals (Level 4)	B	Use knowledge of place value and multiplication facts to 10 × 10 to derive related multiplication and division facts involving decimals (Level 4)	Make and justify estimates and approximations to calculations (Level 5)
Calculating (AT2)					
Add or subtract mentally pairs of two-digit whole numbers (Level 3)	A	Use efficient written methods to add and subtract whole numbers and decimals with up to two places (Level 4)	A	Use efficient written methods to add and subtract integers and decimals, to multiply and divide integers and decimals by a one-digit integer, and to multiply two-digit and three-digit integers by a two-digit integer (Level 5)	Use bracket keys and the memory of a calculator to carry out calculations with more than one step; use the square-root key (Level 5)
Develop and use written methods to record, support and explain multiplication and division of two-digit numbers by a one-digit number, including division with remainders (Level 4)	B				
Understanding shape (AT3)					
Know that angles are measured in degrees and that one whole turn is 360°; compare and order angles less than 180° (Level 3)	A	Read and plot co-ordinates in the first quadrant; recognise parallel and perpendicular lines in grids and shapes; use a set-square and ruler to draw shapes with perpendicular or parallel sides (Level 4)	A	Visualise and draw on grids of different types where a shape will be after reflection, after translations, or after rotation through 90° or 180° about its centre or one of its vertices (Level 5)	Know the sum of angles on a straight line, in a triangle and at a point, and recognise vertically opposite angles (Level 5)
Measuring (AT3)					
Choose and use standard metric units and their abbreviations when estimating, measuring and recording length, weight and capacity; know the meaning of 'kilo', 'centi' and 'milli' and, where appropriate, use decimal notation to record measurements (Level 3)	B	Draw and measure lines to the nearest millimetre; measure and calculate the perimeter of regular and irregular polygons; use the formula for the area of a rectangle to calculate the rectangle's area (Level 4)	A	Select and use standard metric units of measure and convert between units using decimals to two places (Level 4)	Solve problems by measuring, estimating and calculating; measure and calculate using imperial units still in everyday use; know their approximate metric values (Level 5)
Handling data (AT4)					
Answer a question by identifying what data to collect; organise, present, analyse and interpret the data in tables, diagrams, tally charts, pictograms and bar charts, using ICT where appropriate (Level 3)	B	Construct frequency tables, pictograms and bar and line graphs to represent the frequencies of events and changes over time (Level 4)	A	Solve problems by collecting, selecting, processing, presenting and interpreting data, using ICT where appropriate; draw conclusions and identify further questions to ask (Level 5)	Understand and use the probability scale from 0 to 1; find and justify probabilities based on equally likely outcomes in simple contexts (Level 5)

NOTE: **Using and applying mathematics (AT1)** is incorporated throughout

	Foundation Stage	Year 1	Year 2	Year 3	Year 4	Year 5	Year 6
End-of-year National Expectations	1b	1a (2c)	2b	2a (3c)	3b	3a (4c)	4b

Task 1
Using and applying mathematics

Objectives   **NC AT 1**   **NC Level 4**
- Solve one-step and two-step problems involving whole numbers and decimals and all four operations, choosing and using appropriate calculation strategies, including calculator use
- Represent a puzzle or problem by identifying and recording the information or calculations needed to solve it; find possible solutions and confirm them in the context of the problem
- Explain reasoning using diagrams, graphs and text; refine ways of recording using images and symbols

Resources
- RCM 1: Word problem cards (enlarged to A3 and cut out)
- pencil and paper (per child)
- calculator (per child)

Task
- Give each child one of the Word problem cards from RCM 1 and a pencil and a piece of paper. See below for guidance as to which card to give to individual children depending on their ability.

	Easy	**Moderate**	**Difficult**
Cards	1–6	7–14	15–20

> Success criterion: *Read and understand the problem*

- Ask the children to read the card quietly to themselves.
- In turn, go around the group asking each child to explain their word problem to the rest of the group in their own words. Ask: **Amy, what is your problem about? Trevor, what do you have to find out?**

> Success criterion: *Correctly identify which operation(s) to use*

- Ask each child to suggest which operation(s) they need to use to work out the answer to the word problem. Ask: **Matt, which operation(s) do you need to use to work out the answer to your problem?**
- Ask each child to explain how they know which operation to use. Ask: **Matt, how do you know you need to add/subtract/multiply/divide? What clues are there in the problem?**

> Success criterion: *Carry out the calculation(s) to obtain the correct answer using an appropriate method*

- Ask each child to write down the calculation needed to solve the problem and to work out the answer. Say: **On your sheet of paper I want each of you to write down the calculation(s) needed to solve your problem and then I want you to work out the answer.**
- After sufficient time, ask each child to read out the calculation and answer to their problem.
- Encourage each child to talk about the method they used to obtain their answer. Ask: **Trevor, how did you work that calculation out? Show us your working. Is there another way you could have worked it out?**

Success criterion: *Check answer using an effective method*

- Ask each child to check their answer. Say: **I want each of you now to check your answer.**
- After sufficient time ask each child to explain the method they used for checking their answer.

- Repeat the above for the other word problems.
- Conclude by giving each child a calculation, e.g. 481 + 749, £14.12 − £6.52, 16 × 19, 426 ÷ 8, and asking them to make up a word problem that can be solved using the calculation. Say: **Louise, here is your calculation: 481 + 749. Tell us a word problem using this calculation.**

- How many calculations are needed to solve this problem?
- What is the first thing you need to do to work out the answer to this problem?
- Roughly, what is the answer you expect to get from this problem?
- How did you know you needed to add/subtract/multiply/divide? What clues were there in the problem?
- How are you going to record your working?
- What does this answer tell you?
- What are the important things you need to remember when solving word problems?

What to do for those children who achieve *above* expectation

- Use the grid opposite to choose suitable word problem cards.

What to do for those children who achieve *below* expectation

- Use the grid opposite to choose suitable word problem cards.
- Allow the children to use a calculator to help them carry out the calculation needed to obtain the correct answer.

Answers

Question	Problem involving	Number of steps required Operation(s) required	Answer	Question	Problem involving	Number of steps required Operation(s) required	Answer
1.	Measures: Capacity	1 step: multiplication	17 l 500 ml	11.	Money	2 steps: multiplication and subtraction	£90
2.	Real life	1 step: addition	980	12.	Money	1 step: subtraction	£7702.20
3.	Real life	2 steps: multiplication and division	7	13.	Real life	2 steps: division and multiplication	42
4.	Measures: Length	2 steps: multiplication and division	6 km	14.	Measures: Perimeter	1 step: addition or 2 steps: addition and multiplication	55 m
5.	Money	1 step: multiplication	£6.72	15.	Measures: Weight	2 steps: multiplication and division	10 400
6.	Measures: Length and Time	1 step: division	2·2 sec	16.	Measures: Area	3 steps: multiplication, multiplication and addition	396 m²
7.	Measures: Time	2 steps: addition and addition	2 hr 5 min	17.	Real life	3 steps: addition, division and subtraction	2
8.	Measures: Capacity	2 steps: multiplication and subtraction	8 l 200 ml	18.	Money	2 steps: subtraction and division	£6.50
9.	Money	5 steps: division, division, subtraction, division and subtraction	Charlie: £15 Alex: £22.50	19.	Measures: Time	3 steps: addition, addition and subtraction or 3 steps: addition, addition and addition	5 hr 50 min
10.	Measures: Weight	1 step: addition	11 kg 900 g	20.	Real life	2 steps: division and multiplication	100

Using and applying mathematics

Objective	NC AT 1	NC Level 4

- Plan and pursue an enquiry; present evidence by collecting, organising and interpreting information; suggest extensions to the enquiry

See Task 27
Handling data and Using and applying mathematics
Page 68

Task 2
Using and applying mathematics

Objectives   NC AT 1   NC Level 4

- Explore patterns, properties and relationships and propose a general statement involving numbers or shapes; identify examples for which the statement is true or false
- Represent a puzzle or problem by identifying and recording the information or calculations needed to solve it; find possible solutions and confirm them in the context of the problem
- Explain reasoning using diagrams, graphs and text; refine ways of recording using images and symbols

Resources
- RCM 2: Puzzles 1
- RCM 3: Puzzles 2
- pencil and paper (per child)
- calculator – optional (per child)

Task
- Prior to the task, decide which puzzle/investigation to give individual children from RCM 2 and RCM 3. Alternatively, use puzzles or investigations of your own.

> Success criteria: *Explore patterns, properties and relationships*
> *Make a general statement*
> *Solve a problem or puzzle*
> *Explain reasoning*

- Give each child a puzzle/investigation from RCM 2 or RCM 3 and a pencil and a piece of paper.
- Ask each child to read through their puzzle/investigation. Ask: **Alpesh, what is your puzzle/ investigation about? What do you have to find out? What do you know already that can help you solve this?**
- Briefly discuss the puzzle/investigation with each child.
- Say: **I now want each of you to work on your puzzle/investigation. If you need anything, or are unsure of something just ask me.**
- Allow the children sufficient time to spend on their puzzle/investigation. As the children work through the task, ask specific questions to help individual children with the task as well as to assess children's ability to interpret and complete the task.
- Once each child has completed their task, ask each child to talk about what they did and what they found out.
- Encourage each child to talk about any expressions or formulae they constructed. Ask: **Can you tell me the rule you discovered? Did you write a formula? Is your formula in words or symbols? Does your formula always work? How can you be sure?**
- Finally, ask each child to justify why they worked the way they did. Encourage them to explain their methods of working and recording. Ask: **Why did you…? How else could you have gone about it? What did you find easy/difficult about what you did? If you had to do this puzzle/investigation again, how would you do it differently next time?**

 ● What do you have to do in this puzzle/investigation?
● Explain to me what you have discovered.
● Write about what you discovered. Can you explain this using symbols instead of words?

What to do for those children who achieve *above* expectation

● Provide more challenging puzzles/investigations.
● Encourage the children to construct expressions and formulae in symbols.

What to do for those children who achieve *below* expectation

● Provide extra support as the children work through the puzzle/investigation.
● Allow the children to use a calculator.
● Choose easier puzzles/investigations.

Answers

RCM 2 Top

The following statements are true:

● The sum of three even numbers is always even.
● The difference between one odd and one even number is always odd.
● The product of two even numbers is always even.
● The product of two odd numbers is always odd.

The following statements are false (with the correct statement given in brackets):

● The sum of three odd numbers is always even. (The sum of three odd numbers is always odd.)
● The difference between two odd or two even numbers is always odd. (The difference between two odd or two even numbers is always even.)
● The product of one odd and one even number is always odd. (The product of one odd and one even number is always even.)

RCM 2 Bottom

The following statements are true:

● The number 162 is divisible by 3. (A number is divisible by 3 if the sum of the digits is divisible by 3.)
● The number 155 is divisible by 5. (A number is divisible by 5 if the last digit is 5 or 0.)

The following statements are false:

● The number 146 is divisible by 4. (A number is divisible by 4 if the last two digits are divisible by 4.)
● The number 158 is divisible by 6. (A number is divisible by 6 if the number is even and divisible by 3.)

RCM 3 Top

9.44 a.m.

RCM 3 Bottom

20

Task 3
Counting and understanding number

Objective   NC AT 2   NC Level 4

- Count from any given number in whole-number and decimal steps, extending beyond zero when counting backwards; relate the numbers to their position on a number line

---

Resources

- RCM 4: Two-digit number cards
- RCM 5: Three-digit number cards
- RCM 6: Four-digit number cards
- RCM 7: Decimal cards – tenths
- RCM 8: Decimal cards – hundredths
- large sheet of paper and marker
- pencil and paper (per child)

Task

- Prior to the task, shuffle together the cards from RCMs 4–6 and place them face down in a pile. Also shuffle together the cards from RCMs 7 and 8 and place them face down in a separate pile.
- Arrange the children in the group in a circle or semi-circle around the table.

> Success criterion: *Identify the rule and continue the sequence*

- Turn over one of the number cards from RCMs 4–6 and say the number to the children, e.g. **604**.
- Count on or back in steps, e.g. say: **604, 624, 644, 664, 684, 704.**
- Ask: **What is the rule in my sequence?** (+ 20) **Max, continue this number sequence for me.**
- Repeat the above until each child has sufficiently demonstrated their ability to identify and continue a sequence involving whole numbers.
- Repeat the above for number sequences involving decimals with up to two places, using cards from RCMs 7 and 8.

> Success criterion: *Count on in whole number steps*

- Turn over one of the number cards from RCMs 4–6 and ask: **Martin, what is this number?**
- Say: **I want you to count on in steps of three from this number until I say stop. Ready? Go!**
- Continue until the child has continued the count for at least 12 steps.
- Repeat the above until each child has sufficiently demonstrated their ability to count on in whole number steps from any two-digit, three-digit or four-digit number.

> Success criterion: *Count back in whole number steps*

- Turn over one of the number cards from RCMs 4–6 and ask: **Helen, what is this number?**
- Say: **I want you to count back in steps of four from this number until I say stop. Ready? Go!**
- Continue until the child has continued the count for at least 12 steps.
- Repeat the above until each child has sufficiently demonstrated their ability to count back in whole number steps from any two-digit, three-digit or four-digit number, including counting beyond zero when counting backwards.

> Success criterion: *Count on in decimal steps*

- Turn over one of the decimal cards from RCMs 7 and 8 and ask: **Max, what is this number?**
- Say: **I want you to count on in steps of zero point two from this number until I say stop. Ready? Go!**

● Continue until the child has continued the count for at least 12 steps.

● Repeat the above until each child has sufficiently demonstrated their ability to count on in decimal steps from any decimal with up to two places.

> Success criterion: *Count back in decimal steps*

● Turn over one of the decimal cards from RCMs 7 and 8 and ask: **Charlotte, say this number for me.**

● Say: **I want you to count back in steps of zero point five from this number until I say stop. Ready? Go!**

● Continue until the child has continued the count for at least 12 steps.

● Repeat the above until each child has sufficiently demonstrated their ability to count back in decimal steps from any decimal with up to two places, including counting beyond zero when counting backwards.

> Success criterion: *Relate numbers to their position on a number line*

● Choose three or four number or decimal cards from RCMs 4–8 that make part of a number sequence. Arrange the cards on the large sheet of paper as though they appear on a number line, e.g.

● Ask the children to copy the number line and write in the missing numbers.

● Ask: **How did you know what numbers to write on the line?**

● Repeat the above until the children have sufficiently demonstrated their ability to relate numbers in a number sequence to their position on a number line.

● What is the next number in this sequence? How do you know this is the next number?
● What is the rule for this sequence?
● What is the missing number in this sequence?
● Create a sequence that involves the number 1·4. … –5. Describe your sequence.

What to do for those children who achieve *above* expectation

● When asking the children to count on or back in whole number steps ask them to count in two-digit steps, e.g. 12, 15 or 25.

● When asking the children to count on or back in decimal steps ask them to count in steps such as 0·3, 0·9, 1·1, 1·3 or 1·7.

What to do for those children who achieve *below* expectation

● When asking the children to count on or back in whole number steps:
  – only use RCM 4: Two-digit number cards or RCM 5: Three-digit number cards;
  – ask the children to count on or back in single-digit steps or multiples of 10.

● When asking the children to count on or back in decimal steps:
  – use RCM 4: Two-digit number cards;
  – only use RCM 7: Decimal cards – tenths;
  – ask the children to count on or back in simple steps such as: 0·1, 0·2 or 0·5.

Task 4
Counting and understanding number

Objective   NC AT 2   NC Level 4
- **Explain what each digit represents in whole numbers and decimals with up to two places, and partition, round and order these numbers**

Resources
- RCM 4: Two-digit number cards
- RCM 5: Three-digit number cards
- RCM 6: Four-digit number cards
- RCM 7: Decimal cards – tenths
- RCM 8: Decimal cards – hundredths
- large sheet of paper
- marker

Task

NOTE: This task involves quite a number of Success criteria. It is advisable to choose no more than three or four criteria at a time, perhaps focusing on a specific aspect, e.g. place value, ordering or rounding; or either whole numbers or decimals.

- Prior to the task:
  - shuffle together the cards from RCMs 4–6 and place them face down in a pile
  - shuffle together the cards from RCMs 7 and 8 and place them face down in a separate pile
  - on the large sheet of paper, write several whole numbers with five digits, six digits or seven digits, e.g. 4 768 301, 34 298, 540 679, 87 645, 762 081, 1 820 346…
- Arrange the children in the group in a circle or semi-circle around the table.

  Success criterion: *Know what each digit in a whole number represents*

- Turn over one of the number cards from RCMs 4–6 and pointing to one of the digits in the number, ask: **Michael, what does this digit represent?**
- Repeat several times for each child.
- Referring to a different number card for each child, say: **John, point to the digit that shows how many hundreds/units/tens are in this number.**
- Next, referring to the numbers on the large sheet of paper, repeat the above asking the children to explain what each digit represents in a five-digit, six-digit or seven-digit number.

  Success criterion: *Order a mixed set of whole numbers*

- Collect and shuffle all the number cards from RCMs 4–6.
- Lay five cards, face up, in front of each child. Say: **Look at the numbers in front of you. I want each of you to place these cards in order, smallest to largest.**
- Once each child has done this give each child another card and say: **Look at the cards you have just put in order. Where would you put this number so that the order is still correct?**
- Repeat the above until each child has sufficiently demonstrated their ability to order a mixed set of whole numbers to 10 000.

● Then referring to the numbers written on the large sheet of paper, ask the children, as a group, to order these numbers from the smallest to the largest. Ask: **Which is the smallest of these numbers? How do you know? What is the largest number? Let's order these numbers from smallest to largest. If … is the smallest number, which number comes next?**

> Success criteria: *Round a two-digit, three-digit or four-digit whole number to the nearest 10*
> *Round a three-digit or four-digit whole number to the nearest 100*
> *Round a four-digit whole number to the nearest 1000*

● Distribute the number cards from RCMs 4–6 amongst the children.

● Ask the children to place their cards face down in a pile in front of them.

● In turn, ask each child to turn over the top card from their pile and to round their number to the nearest 10, 100 and/or 1000, depending on the number that they have turned over. Ask: **Louise, what is 458 rounded to the nearest 10? What is 458 rounded to the nearest 100?**

● Continue until each child has sufficiently demonstrated their ability to round a two-digit, three-digit or four-digit number to the nearest 10, 100 and/or 1000.

● Place the number cards from RCMs 4–6 to one side.

> Success criterion: *Know what each digit in a decimal with up to two places represents*

● Turn over one of the decimal cards from RCMs 7 and 8 and, pointing to one of the digits in the number, ask: **Louise, what does this digit represent? What is the value of the 5 in this number?**

● Repeat several times for each child.

● Referring to a different decimal card for each child, ask: **Michael, point to the digit that shows how many units/tenths/hundredths are in this number.**

● Repeat the above until each child has sufficiently demonstrated their ability to partition decimals with up to two places.

> Success criterion: *Order a mixed set of decimals with up to two places*

● Collect and shuffle all the number cards from RCMs 7 and 8.

● Lay five cards, face up, in front of each child. Say: **Look at the numbers in front of you. I want each of you to place these cards in order, smallest to largest.**

● Once each child has done this give each child another card and say: **Look at the cards you have just put in order. Where would you put this number so that the order is still correct?**

● Repeat the above until each child has sufficiently demonstrated their ability to order a mixed set of decimals with up to two places.

> Success criteria: *Round a decimal with up to two places to the nearest whole number*
> *Round a decimal with two places to the nearest tenth*

● Distribute the number cards from RCMs 7 and 8 amongst the children.

● Ask the children to place their cards face down in a pile in front of them.

● In turn, ask each child to turn over the top card from their pile and to round their number to the nearest whole number and/or tenth, depending on the number that they have turned over. Ask: **Louise, what is 4·6 rounded to the nearest whole number? Michael what is 5·35 rounded to the nearest whole number? What is 5·35 rounded to the nearest tenth?**

● Continue until each child has sufficiently demonstrated their ability to round a decimal with up to two places to the nearest whole number and/or tenth.

● What does the digit 6 represent in the number 9·36? What about the 3? …9?
● What if it was a length given in metres? …a mass given in kilograms? …a capacity in litres?
● What is 3·91 rounded to the nearest whole number? …rounded to the nearest tenth?

What to do for those children who achieve *above* expectation

● Ask the children to partition, round and order any whole number.
● Ask the children to partition, round and order decimals with up to three places.

What to do for those children who achieve *below* expectation

● Ask the children to partition, round and order whole numbers with up to four-digits.
● Ask the children to partition, round and order decimals with one place.

Task 5
Counting and understanding number

Objective   NC AT 2 NC Level 4

- Express a smaller whole number as a fraction of a larger one, e.g. recognise that 5 out of 8 is $\frac{5}{8}$; find equivalent fractions, e.g. $\frac{7}{10} = \frac{14}{20}$, or $\frac{19}{10} = 1\frac{9}{10}$; relate fractions to their decimal representations

Resources

- 1 set of 0–9 digit cards
- RCM 9: Equivalent fraction cards
- RCM 10: Relating fractions to their decimal representations 1 (per pair)
- RCM 11: Relating fractions to their decimal representations 2 (enlarged to A3)
  (for children achieving *above* and *below* expectation)
- coloured pencil (per child)
- large sheet of paper and marker

Task

> Success criterion: *Express a smaller whole number as a fraction of a larger one*

- Shuffle the set of 0–9 digit cards and place them face down in a pile in the middle of the table.
- Turn over two of the cards, e.g. 3 and 7, then ask questions similar to the following: **Leo, what fraction of 7 is 3?** ($\frac{3}{7}$) **Leo, 3 is what fraction of 7?** ($\frac{3}{7}$) or **Leo, I had 7 sweets and I ate 3 of them. What fraction of the sweets did I eat?** ($\frac{3}{7}$)
- Repeat the above until each child has sufficiently demonstrated their ability to express a smaller whole number as a fraction of a larger one.

> Success criterion: *Find equivalent fractions*

- Spread the Equivalent fraction cards from RCM 9 face up in a random order on the table.
- Pointing to the $\frac{1}{2}$ card, ask a child to identify an equivalent fraction, say: **Natasha, point to a card on the table that is equivalent to a half.**
- Ask two other children to point to the two other cards that are equivalent to a $\frac{1}{2}$ (i.e. $\frac{1}{2} = \frac{2}{4} = \frac{3}{6} = \frac{4}{8}$).
- Ask children to suggest other fractions, that are not on the cards, that are equivalent to a half. Say: **Leo, tell me another fraction that is equivalent to a half.**
- Repeat, pointing to other cards and asking a child to identify equivalent fractions,

 $\frac{1}{3} = \frac{2}{6} = \frac{3}{9} = \frac{4}{12}$ $\frac{1}{5} = \frac{2}{10} = \frac{3}{15} = \frac{4}{20}$ $\frac{3}{4} = \frac{6}{8}$

 $\frac{1}{4} = \frac{2}{8} = \frac{3}{12} = \frac{4}{16}$ $\frac{1}{6} = \frac{2}{12} = \frac{3}{18} = \frac{4}{24}$ $\frac{2}{3} = \frac{4}{6}$

- Conclude by writing fractions similar to the following onto the large sheet of paper and asking children to suggest equivalent fractions:

 – unitary fractions: $\frac{1}{7}$, $\frac{1}{8}$, $\frac{1}{9}$, $\frac{1}{12}$

 – non-unitary fractions: $\frac{3}{4}$, $\frac{2}{3}$

 – tenths: $\frac{3}{10}$, $\frac{7}{10}$, $\frac{9}{10}$

 – hundredths: $\frac{2}{100}$, $\frac{14}{100}$

 – improper fractions: $\frac{12}{10}$, $\frac{7}{3}$

 – mixed numbers: $1\frac{7}{10}$, $2\frac{2}{3}$

> Success criterion: *Relate fractions to their decimal representations*

- Arrange the children into pairs.
- Provide each pair with a copy of RCM 10 and a coloured pencil each. Ensure that children in the same pair do not have the same colour pencil.
- Explain to the children that they are going to play a game that involves matching fractions and their decimal equivalents.
- Explain the rules to the children:
  - Children take turns to choose two numbers on the acrobats' shirts and make a fraction using one number as the numerator and the other as the denominator, e.g. $\frac{3}{4}$.
  - The child whose turn it is then identifies the decimal equivalent of their fraction on the stand of the paper hoop (e.g. 0·75) and writes the fraction inside the star using their coloured pencil, i.e. $\frac{3}{4}$.
  - Each fraction can only be made once, however equivalent fractions, e.g. $\frac{1}{2}$, $\frac{2}{4}$, $\frac{4}{8}$, can all be made and the equivalent fraction written in the same hoop.
  - If the decimal is not on the stand of a paper hoop then the child misses a turn.
  - The winner is the child with more fractions written inside the stars in their colour.
- As the children play the game, assess each child's ability to relate fractions to their decimal representations.

- What fraction of 5 is 2? 2 is what fraction of 5?
- Tell me some fractions that are equivalent to a quarter. How do you know they are equivalent to a quarter? Are there any others?
- Tell me a fraction that is the same as 0·5. Can you think of another one?
- Tell me the decimal equivalent of a quarter, a third, a tenth…

What to do for those children who achieve *above* expectation

- Find equivalent fractions: Ask the children to convert improper fractions to mixed numbers and vice versa, e.g. ask questions similar to the following:
  - **Leo, change $\frac{20}{17}$ into a mixed number.**
  - **Natasha, what is $4\frac{2}{3}$ as an improper fraction?**
- Relate fractions to their decimal representation: Play the game as above using the top game board on RCM 11.

What to do for those children who achieve *below* expectation

- Express a smaller whole number as a fraction of a larger one: Present children with real life situations, e.g. say: **I had 7 sweets and I ate 3 of them. What fraction of the sweets did I eat?**
- Find equivalent fractions: Ask the children to find the equivalence of unitary fractions and tenths only.
- Relate fractions to their decimal representation: Play the game as above using the bottom game board on RCM 11.

Task 6
Counting and understanding number

Objective NC AT 2 NC Level 4
• Understand percentage as the number of parts in every 100 and express tenths and hundredths as percentages

Resources
● RCM 12: Percentages 1 (enlarged to A3)
● large sheet of paper and marker

Task
● Prior to the task, on the large sheet of paper, write some fractions and decimals involving tenths and hundredths, e.g. $\frac{7}{10}$, $\frac{3}{10}$, 0·1, 0·5, $\frac{18}{100}$, 0·34, $\frac{28}{100}$, 0·92, $\frac{62}{100}$…

> Success criterion: *Understand percentage as the number of parts in every 100*

● Show the children RCM 12.
● Pointing to specific shapes on the RCM, ask individual children to say what percentage of the shape is shaded.
● Ask: **Lisa, what percentage of this shape is shaded?**
● Repeat asking children to say what percentage of the shape is not shaded.
● Ask: **David, what percentage of this shape is not shaded?**
● Continue until each child has sufficiently demonstrated their understanding of percentage as the number of parts in every 100.
● Place RCM 12 to one side.

> Success criterion: *Express tenths and hundredths as percentages*

● Referring to the large sheet of paper showing tenths and hundredths, point to one of the fractions/decimals and ask: **Lisa, what is zero point five as a percentage? David, what is this fraction as a percentage?**
● Continue until each child has sufficiently demonstrated their ability to express tenths and hundredths as a percentage.

● What does the term 'per cent' mean?
● What is $\frac{23}{100}$ as a percentage? What is $\frac{7}{10}$ as a percentage?

What to do for those children who achieve *above* expectation
● Referring to RCM 12, ask the children to express Shapes 9 to 13, 15, 17 and 18 as percentages (shaded and not shaded).

What to do for those children who achieve *below* expectation

● Referring to RCM 12, only ask the children to express Shapes 1 to 8, 14 and 16 as percentages (shaded and not shaded).

Answers

Shape	% of shape shaded	% of shape not shaded
1	60%	40%
2	70%	30%
3	25%	75%
4	50%	50%
5	100%	0%
6	64%	36%
7	47%	53%
8	30%	70%
9	40%	60%
10	60%	40%
11	75%	25%
12	25%	75%
13	50%	50%
14	90%	10%
15	75%	25%
16	0%	100%
17	80%	20%
18	20%	80%

Task 7
Counting and understanding number

Objective   NC AT 2   NC Level 4
- Use sequences to scale numbers up or down; solve problems involving proportions of quantities, e.g. decrease quantities in a recipe designed to feed six people

Resources
- RCM 13: Proportion 1 (enlarged to A3)
- RCM 14: Proportion 2 (enlarged to A3 and cut out)

Task

> Success criterion: *Understand the concept of proportion*

- Show the children RCM 13.
- Explain to the children that this RCM shows 16 different wall tile patterns.
- Ensure that the children are familiar with the terms floral tiles and plain tiles.
- Pointing to the first wall tile pattern ask a child to name the proportion of floral tiles in the whole pattern.
- Ask: **Jake, look at the first wall tile pattern. As a fraction, what is the proportion of floral tiles in the whole pattern?**
- Repeat the above for the remaining patterns on the RCM.
- Occasionally ask the children to explain their method of working. Say: **Sarah, how do you know that the proportion of floral tiles in this pattern is $\frac{1}{5}$?**
- As you go through the task with the children, assess each child's understanding of the concept of proportion.
- Place RCM 13 to one side.

> Success criteria: *Use sequences to scale numbers up or down*
> *Solve problems involving proportions of quantities*

- Show the children one of the recipes from RCM 14.
- Discuss the recipe with the children, highlighting the number of people the recipe is for and the quantities for the different ingredients. Ask questions such as: **How many different ingredients are there in this recipe? This recipe is for how many people? How many grams of chicken do you need to serve two people? ...cloves of garlic...?**
- Then ask the children to scale the recipe up (or down) for a larger (or smaller) number of people. Ask questions such as: **How many grams of noodles would you need if you were to make this recipe for 4 people? ...6 people? ...3 people? How many millilitres of oyster sauce would you need if you were going to make this recipe to feed 10 people? If someone was making this recipe just for themselves, how many grams of chicken would they use? How many grams would you need to feed twice as many people?**
- Repeat the above for the other recipes, until each child has sufficiently demonstrated their ability to scale numbers up or down when solving problems involving proportions of quantities.

- How did you find the proportion of ... compared to the whole?
- How many millilitres would you need to feed twice as many people?
- This recipe feeds 4 people. How many grams would you need to feed 8 people? ...2 people? ...6 people?

What to do for those children who achieve *above* expectation

- Referring to RCM 13, ask the children to describe the ratio of floral tiles to plain tiles (Level 5)

What to do for those children who achieve *below* expectation

- Referring to RCM 14, only ask the children questions that involve doubling or halving the quantities.

Answers

RCM 13: Proportion 1

Pattern	Proportion of floral tiles in whole pattern	Ratio of floral tiles to plain tiles
1	$\frac{1}{2}$	1 : 1
2	$\frac{1}{5}$	1 : 4
3	$\frac{2}{3}$	2 : 1
4	$\frac{1}{4}$	1 : 3
5	$\frac{3}{5}$	3 : 2
6	$\frac{2}{3}$	2 : 1
7	$\frac{4}{5}$	4 : 1
8	$\frac{2}{5}$	2 : 3
9	$\frac{2}{5}$	2 : 3
10	$\frac{1}{3}$	1 : 2
11	$\frac{4}{5}$	4 : 1
12	$\frac{1}{2}$	1 : 1
13	$\frac{1}{5}$	1 : 4
14	$\frac{3}{5}$	3 : 2
15	$\frac{1}{4}$	1 : 3
16	$\frac{2}{5}$	2 : 3

Task 8
Knowing and using number facts

Objective NC AT 2 NC Level 4
- **Use knowledge of place value and addition and subtraction of two-digit numbers to derive sums and differences and doubles and halves of decimals, e.g. 6·5 ± 2·7, halve 5·6, double 0·34**

Resources
- large sheet of paper and marker
- RCM 4: Two-digit number cards
- RCM 7: Decimal cards – tenths
- RCM 8: Decimal cards – hundredths (for children achieving *above* expectations)
- pencil and paper (optional, per child)

Task
- Prior to the task:
  - on the large sheet of paper, write some two-digit decimals involving hundredths, e.g. 0·56, 0·74, 0·32, 0·49, 0·28… Be sure to include several that include an even number of hundredths.
  - shuffle and place each set of number cards into a pile of its own.

> Success criterion: *Add and subtract any pair of two-digit whole numbers*

- Using the two-digit number cards from RCM 4, place two cards face up in front of each child. Ask questions such as: **What is the total of these two numbers? Find the sum of these two numbers. Add these numbers together.**
- Then ask questions such as: **What is the difference between these two numbers? Subtract the smaller number from the larger number.**
- Continue until each child has sufficiently demonstrated their ability to add and subtract any pair of two-digit whole numbers.

> Success criterion: *Add and subtract any pair of two-digit decimals*

- Using the tenths decimal cards from RCM 7, repeat the above asking the children to add, then subtract a pair of two-digit decimals involving units and tenths.
- Referring to the large sheet of paper showing two-digit decimals involving hundredths, e.g. 0·56, 0·74, point to two of the decimals and ask a child to add, then subtract the two numbers.
- Occasionally ask questions such as: **How do you know that answer? How did you work it out? What number fact did you use to work out that answer?**
- Continue until each child has sufficiently demonstrated their ability to add and subtract any pair of two-digit decimals involving units and tenths, e.g. 1·5, 4·6; and hundredths, e.g. 0·56, 0·74.

> Success criterion: *Double two-digit decimals*

- Using the tenths decimal cards from RCM 7, place one card face up in front of each child. Ask questions such as: **Double this number. What is twice this number? Multiply this number by two.**
- Referring to the large sheet of paper showing two-digit decimals involving hundredths, e.g. 0·56, 0·74, point to one of the decimals and ask a child to double the number.
- Continue until each child has sufficiently demonstrated their ability to double two-digit decimals involving units and tenths, e.g. 1·5, 4·6; and hundredths, e.g. 0·56, 0·74.

Success criterion: *Halve two-digit decimals*

- Using the tenths decimal cards from RCM 7, place one card face up in front of each child. Ask questions such as: **Halve this number. Divide this number by two.**
- Referring to the large sheet of paper showing two-digit decimals involving hundredths, e.g. 0·56, 0·74, point to one of the decimals and ask a child to halve the number.
- Continue until each child has sufficiently demonstrated their ability to halve two-digit decimals involving units and tenths, e.g. 1·5, 4·6; and hundredths, e.g. 0·56, 0·74.

- Add these two numbers together. What is the difference between these two numbers?
- What number added to this number gives 10? ...7·5? ...0·54?
- How did you work out the answer to these calculations? Talk me through your method.
- What addition/subtraction fact did you use to help you work out the answer to this calculation?

What to do for those children who achieve *above* expectation

- Using the hundredths decimal cards from RCM 8, ask the children to add, subtract and double decimals with two places.

What to do for those children who achieve *below* expectation

- Do not ask children to halve a decimal with an odd number of tenths, e.g. 6·9, 9·1, or hundredths, e.g. 0·49, 0·57.

Task 9
Knowing and using number facts

Objective NC AT 2 NC Level 4
- Recall quickly multiplication facts up to 10 × 10 and use them to multiply pairs of multiples of 10 and 100; derive quickly corresponding division facts

Resources
- pencil and paper (per child and yourself)
- 2 × 1–10 or 0–9 dice
- RCM 15: Division facts (enlarged to A3)

Task
- Provide each child with a pencil and a piece of paper. Ask them to write the numbers 1 to 25 down the left-hand side of the sheet. You do the same.

  Success criterion: Recall multiplication facts up to 10 × 10

- Say: **I'm going to roll these two 1–10 (or 0–9) dice and call out the numbers rolled. I then want you to quickly multiply the two numbers together and write the answer beside number one on your sheet of paper. I'm then going to roll the dice again and I want you to multiply the two numbers together and write the answer beside number two on your sheet of paper. Got the idea? I'm going to do this 15 times.**
- Roll the dice and call out the two numbers rolled.
- Go through the task with the children quickening the pace as the children become more confident with the task. Ensure that you also write down the answer as you go along for checking purposes at a later stage.
- Stop after 15 rolls of the dice and when the children have written an answer beside numbers 1 to 15 on their sheet of paper.

  Success criterion: Multiply pairs of multiples of 10 and 100

- Say: **I'm now going to call out another ten multiplication calculations. However this time they involve multiplying pairs of multiples of 10 and 100. Ready?**
- Read the following calculations to the children, including the question number.
16. 50 × 60 =	**17.** 40 × 80 =	**18.** 300 × 20 =	**19.** 700 × 80 =	**20.** 300 × 600 =
21. 900 × 40 =	**22.** 600 × 70 =	**23.** 80 × 300 =	**24.** 400 × 700 =	**25.** 50 × 80 =

  Success criterion: Derive quickly division facts corresponding to the multiplication facts up to 10 × 10

- Show the children RCM 15.
- Point and say: **Look at the planets and the suns. The suns are called Sun 2, Sun 3, Sun 4, Sun 5 and so on. I'm going to point to one of the planets that is around one of the suns and call out one of your names. The person I call out has to divide the number on the planet by the number on the sun. Got the idea?**
- Point to one of the planets around Sun 2, e.g. 16, and say a child's name.
- The child then divides 16 by 2 and calls out the answer.
- Repeat the above until each child has sufficiently demonstrated their ability to derive quickly division facts corresponding to multiplication tables up to 10 × 10.

- What is 8 times / multiplied by / lots of / groups of 6?
- What is the product of 5 and 9?
- Multiply 60 by 20. What is 400 times 50?
- What is 72 divided by / shared between 8?
- How many sevens are there in 42?
- Divide 35 by 7.
- What multiplication fact did you use to help you work out the answer to that division calculation?
- What division facts start with the number 24 / 42 / 56 and have a whole number answer?

What to do for those children who achieve *above* expectation

- Ask the children division facts corresponding to the multiplication of pairs of multiples of 10 and 100, e.g. $56\,000 \div 70 = \square$, $18\,000 \div 60 = \square$, $540\,000 \div 600 = \square$, $35\,000 \div \square = 70$

- Ask the children quick fire questions involving division facts in the algebraic form, e.g. $48 \div \square = 8$, $\square \div 3 = 7$, $8 = 32 \div \square$, $5 = 45 \div \square$.

What to do for those children who achieve *below* expectation

- Only ask the children to answer the questions involving the 2, 3, 4, 5 and 10 multiplication facts and the corresponding division facts.

Answers

16. $50 \times 60 = 3000$
17. $40 \times 80 = 3200$
18. $300 \times 20 = 6000$
19. $700 \times 80 = 56\,000$
20. $300 \times 600 = 180\,000$
21. $900 \times 40 = 36\,000$
22. $600 \times 70 = 42\,000$
23. $80 \times 300 = 24\,000$
24. $400 \times 700 = 280\,000$
25. $50 \times 80 = 4000$

RCM 15: Division facts

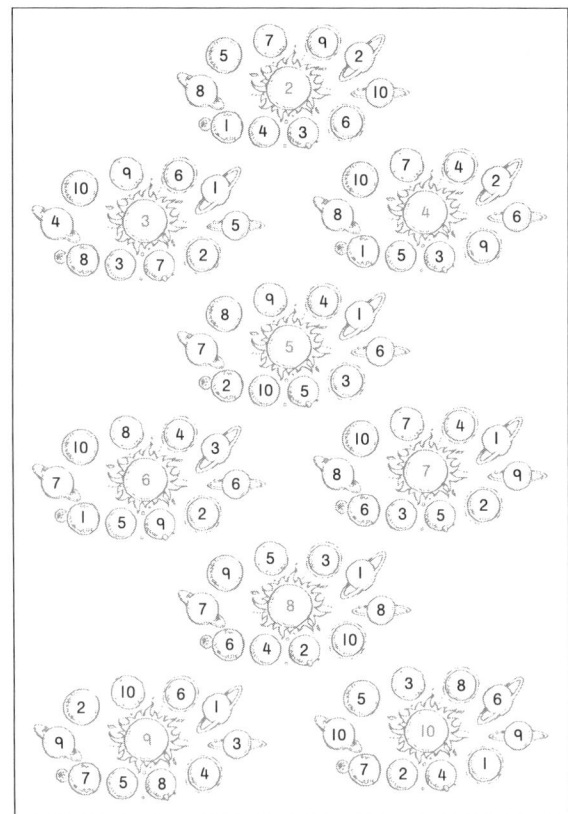

Task 10
Knowing and using number facts

Objective NC AT 2   NC Level 4
- Identify pairs of factors of two-digit whole numbers and find common multiples, e.g. for 6 and 9

Resources
- RCM 4: Two-digit number cards
- RCM 5: Three-digit number cards (for children achieving *above* expectations)
- Multiplication square (for children achieving *below* expectations)
- pencil and paper (per child and yourself)
- 1 set of 0–9 digit cards

Task
- Prior to the task, remove the digits 0 and 1 from the set of 0–9 digit cards. These will not be needed for this task.

> Success criterion: *Identify pairs of factors of two-digit whole numbers*

- Ask: **What is a factor?** (A whole number which will divide exactly into another whole number.) Discuss the various responses offered by the children.
- Say: **I'm going to give each of you a two-digit number card and I want you to write down all the pairs of factors for that number.**
- Demonstrate this to the children by choosing a number card from RCM 4, e.g. 26, and writing down all the pairs of factors for that number, i.e. (1, 26) and (2 and 13).
- Give each child a two-digit number card. As children identify and write down all the factors, ask questions such as: **Is 2 a factor of this number? How do you know? What about 5? What are the factors for your number? Is there another pair of factors? How many factors does your number have?**
- Repeat the above giving each child another number card until each child has sufficiently demonstrated their ability to identify pairs of factors of two-digit whole numbers.

> Success criterion: *Find multiples*

- Ask: **What is a multiple?** (Multiples of any number can be divided exactly by that number.) Discuss the various responses offered by the children.
- Say: **I'm going to give each of you a digit card and I want you to write down as many two-digit multiples of that number as you can.**
- Demonstrate this to the children by choosing a digit card, e.g. 6, and writing down about 6 multiples of that number, i.e. 12, 18, 30, 36, 60, 72, …
- Give each child a digit card. As children write down different multiples, ask questions such as: **Is 16 a multiple of your number? How do you know? What about 34? Tell me some two-digit multiples of your number.**
- Repeat the above giving each child another digit card until each child has sufficiently demonstrated their ability to identify two-digit multiples of numbers 2 to 10.

> Success criterion: *Find common multiples*

● Say: **This time I'm going to give each of you two digit cards and I want you to write down as many common multiples for these two numbers as you can.**
● Demonstrate this to the children by choosing two digit cards, e.g. 3 and 5, and writing down about four common multiples for these numbers, i.e. 15, 30, 45, 60, …
● Give each child two digit cards. As children write down common multiples, ask questions such as: **Is 12 a multiple of …? Is 12 also a multiple of …? How do you know? Is … a common multiple of … and …? How do you know? Tell me a number that is a common multiple of … and … .**
● Repeat the above giving each child two different digit cards until each child has sufficiently demonstrated their ability to identify common multiples of numbers 2 to 10.

● Tell me all the factors of 26.
● How many pairs of factors does the number 71 have? What do we call this number?
● Is 8 a factor of 32? Why? What about 7? Why not?
● Tell me a two-digit multiple of 6. Can you tell me some more?
● Tell me a common multiple of 2 and 5.
● Tell me a number that is both a multiple of 4 and a multiple of 6.

What to do for those children who achieve *above* expectation

● Using RCM 5: Three-digit numbers, ask the children to identify pairs of factors for three-digit numbers.
● Ask the children to find three-digit multiples and common multiples.

What to do for those children who achieve *below* expectation

● Provide children with a multiplication square to assist them in identifying factors and multiples.
● Ask children to find multiples of 2, 3, 4 or 5 only.
● Do not ask the children to find common multiples.

Answers

RCM 4: 2-digit number cards

Number	Pairs of factors	Number	Pairs of factors
14	(1, 14) (2, 7)	17	(1, 17)
23	(1, 23)	26	(1, 26) (2, 13)
32	(1, 32) (2, 16) (4, 8)	38	(1, 38) (2, 19)
41	(1, 41)	45	(1, 45) (3, 15) (5, 9)
50	(1, 50) (2, 25) (5, 10)	59	(1, 59)
62	(1, 62) (2, 31)	64	(1, 64) (2, 32) (4, 16) (8, 8)
67	(1, 67)	71	(1, 71)
75	(1, 75) (3, 25) (5, 15)	76	(1, 76) (2, 38) (4, 19)
80	(1, 80) (2, 40) (4, 20) (5, 16) (8, 10)	89	(1, 89)
93	(1, 93) (3, 31)	96	(1, 96) (2, 48) (3, 32) (4, 24) (6, 16) (8, 12)

Task 11
Knowing and using number facts

Objective	NC AT 2	NC Level 4

• Use knowledge of rounding, place value, number facts and inverse operations to estimate and check calculations

Resources
● RCM 16: Estimate, calculate and check (per child)
● pencil (per child)

Task

NOTE: Prior to the task, write from four to six different calculations in the first column of RCM 16. The calculations should include addition, subtraction, multiplication and division, and be appropriate to individual children's ability:
– add a pair of two-digit numbers
– add a pair of two digit decimals, e.g. 7·8 + 4·6, 0·34 + 0·59
– subtract a pair of two-digit numbers
– subtract a pair of two digit decimals, e.g. 3·5 − 2·9, 0·53 − 0·28
– multiply a two-digit number by a one-digit number
– divide a two-digit number by a one-digit number.

> Success criteria: *Estimate results using a variety of strategies*
> *Check results using a variety of strategies*

● Ask the children to look at the calculations on their sheet.
● Tell the children that for each calculation you want them to:
– estimate the answer first, writing their estimate and any working out in the second column
– work out the answer, showing all their working in the third column
– check their answer, writing any working out in the fourth column.
● As the children work through the calculations on the sheet, ask them questions similar to those below, assessing their ability to use rounding, place value, number facts and inverse operations to estimate and check results.

● What is the approximate answer to this calculation? How did you make your estimate?
● What is the answer to this calculation? How did you work it out?
● How close is the actual answer to your estimate? So do you think that your answer is right? Why?
● Is your answer correct? How can you be so sure?

What to do for those children who achieve *above* expectation
● Include calculations that involve written addition and subtraction calculations involving whole numbers and decimals with up to two places, and multiplication and division calculations such as HTU × U, TU × TU, U·t × U and HTU ÷ U.

What to do for those children who achieve *below* expectation
● Do not include calculations involving decimals.

Task 12
Calculating

Objective   NC AT 2   NC Level 4

- Extend mental methods for whole-number calculations, e.g. to multiply a two-digit by a one-digit number (e.g. 12 × 9), to multiply by 25 (e.g. 16 × 25), to subtract one near multiple of 1000 from another (e.g. 6070 − 4097)

Resources

- RCM 4: Two-digit number cards
- 1 set of 0–9 digit cards
- RCM 17: Calculating a difference mentally (enlarged to A3)
- 2 small counters
- pencil and paper (for children achieving *below* expectations)

Task

- Prior to the task, remove the digits 0 and 1 from the set of 0–9 digit cards. These will not be needed for this task.

  > Success criterion: *Multiply a two-digit number by a one-digit number*

- Place a two-digit number card from RCM 4 and a digit card face up in front of each child. Ask questions such as: **What is the product of these two numbers? Multiply these two numbers together.**
- Occasionally ask questions such as: **How did you work it out? How did you get that answer? How else could you have worked out the answer?**
- Continue until each child has sufficiently demonstrated their ability to multiply a two-digit number by a one-digit number.

  > Success criterion: *Divide a two-digit number by a one-digit number*

- Repeat the above, asking the children to divide the two-digit number by the one-digit number.

  > Success criterion: *Multiply a two-digit number by 50 or 25*

- Place a two-digit number card from RCM 4 in front of each child. Say: **Multiply this number by 50. ...25.**
- Occasionally ask questions such as: **How did you work it out? How did you get that answer? How else could you have worked out the answer?**
- Continue until each child has sufficiently demonstrated their ability to multiply a two-digit number by 50 or 25.

  > Success criterion: *Subtract one near multiple of 1000 from another*

- Place RCM 17 on the table in front of the children.
- Place one of the counters on one of the computers and the other counter on one of the calculators.
- Ask: **Trevor, what is the difference between these two numbers? Can you find it on the notice board?**

- Occasionally ask the children to explain their method of working, ask: **Trevor, how did you work that out?**
- Repeat the above several times for each child.
- Ask the children to choose a number on a computer screen and a number on a calculator and find the difference. Say: **Place one of the counters on one of the computers and the other counter on one of the calculators.** Ask: **What is the difference between these two numbers?**
- Continue until each child has sufficiently demonstrated their ability to calculate mentally a difference such as 6998 − 4009.

- How did you work out the answer to this calculation? Talk me through your method.
- How would you work out this answer by counting on? How would you work out this answer by counting back? Which do you find easier? Why?
- Tell me a subtraction calculation that will give the answer 4002.

What to do for those children who achieve *above* expectation

- Carefully choose which numbers you ask the children to find the difference between, e.g. numbers that cross several multiples of 1000 boundaries, i.e. 7005 − 3994, 6002 − 3994, 6998 − 2993, 8006 − 2993, 8006 − 3997, 5992 − 1999.

What to do for those children who achieve *below* expectation

- Do not ask the children to divide a two-digit number by a one-digit number where the answer has a remainder.
- Carefully choose which numbers you ask the children to find the difference between, e.g. numbers that cross only one multiple of 1000 boundary, i.e. 4001 − 3997, 4009 − 4001, 6002 − 5992, 2009 − 1999, 7005 − 6998, 8006 − 7005.
- Allow the children to use pencil and paper to record their working out.

Task 13
Calculating
(Knowing and using number facts)

Objectives   NC AT 2   NC Level 4
- **Use efficient written methods to add and subtract whole numbers and decimals with up to two places**
- Use knowledge of rounding, place value, number facts and inverse operations to estimate and check calculations (optional)

Resources

- RCM 4: Two-digit number cards
- RCM 5: Three-digit number cards
- RCM 6: Four-digit number cards
- RCM 7: Decimal cards – tenths
- RCM 8: Decimal cards – hundredths
- 0–9 die
- pencil and paper or RCM 16: Estimate, calculate and check (per child)

Task

NOTE: This task involves a number of Success criteria. It is advisable to choose no more than three or four criteria at a time.

- Prior to the task, shuffle and place each set of number cards in a pile of its own.
- Provide each child with a pencil and a piece of paper or RCM 16: Estimate, calculate and check. If you decide to use this RCM, ask the children to estimate and check each calculation.

> Success criterion: *Use efficient written methods to add whole numbers*

- Using the three-digit number cards from RCM 5, choose two cards and place them in front of each child. Say: **Using the pencil and paper I want each of you to add the two numbers together using a written method.**
- Allow the children sufficient time to complete their calculation then ask: **Stanley, what is your answer? How did you get your answer? Gopal, what's your answer? How did you work it out?**
- Collect and reshuffle the cards.
- Repeat the above, using the two-digit, three-digit and four-digit number cards from RCMs 4, 5 and 6, asking the children to use efficient written methods to solve calculations involving:

    ThHTU + HTU

    ThHTU + ThHTU

    addition of more than two numbers

- If appropriate, repeat the above until each child has sufficiently demonstrated their ability to carry out an efficient written method to add two (or more) whole numbers.

> Success criterion: *Use efficient written methods to subtract whole numbers*

- Repeat the above, using the three-digit and four-digit number cards from RCMs 5 and 6, asking the children to use efficient written methods to solve calculations involving:

    HTU − HTU

    ThHTU − HTU

    ThHTU − ThHTU

> **Success criterion:** *Use efficient written methods to add decimals*

● Repeat the above, using the tenths decimal cards from RCM 7 and hundredths decimal cards from RCM 8 to ask the children to use efficient written methods to add two (or more) decimals.

> **Success criterion:** *Use efficient written methods to subtract decimals*

● Repeat the above, using the tenths decimal cards from RCM 7 and hundredths decimal cards from RCM 8 to ask the children to use efficient written methods to subtract decimals.

● What is the answer to that calculation? How did you work it out?
● Did you make an estimate of what the answer might be first? How did you do it?
● Could you have worked out that calculation using a different method? How?
● Did you check your answer? What did you do?

What to do for those children who achieve *above* expectation

● Using a selection of number cards from RCMs 4–8, choose several cards and place them in front of each child. Ask the children to add whole numbers and decimals together.

What to do for those children who achieve *below* expectation

● Encourage the children to use an informal pencil and paper method.
● Only ask the children to solve calculations involving HTU ± TU or HTU ± HTU.

Task 14
Calculating

Objective   NC AT 2   NC Levels 4 & 5

- Use understanding of place value to multiply and divide whole numbers (Level 4) and decimals (Level 5) by 10, 100 or 1000

Resources

- RCM 4: Two-digit number cards
- RCM 5: Three-digit number cards
- RCM 6: Four-digit number cards
- RCM 7: Decimal cards – tenths
- RCM 8: Decimal cards – hundredths

Task

- Prior to the task, shuffle together the cards from RCMs 4–6 and place them face down in a pile. Also shuffle together the cards from RCMs 7 and 8 and place them face down in a separate pile.
- Arrange the children in the group in a circle or semi-circle round the table.

> Success criterion: *Multiply a whole number by 10, 100 or 1000*

- Evenly distribute the cards from RCMs 4–6 amongst the children.
- Children place their cards face down in a pile in front of them.
- Say: **We're going to go around the group and take turns to turn over the top card in your pile. I'm then going to say either 10, 100 or 1000 and I want you to multiply the number on your card by the number I call out. Got the idea?**
- Demonstrate this to the children by asking the first child to turn over their card, e.g. 476. Say: **100.** The child then multiplies 476 by 100 and calls out the answer.
- Repeat the above, rotating around the group until the children have used all their cards or until each child has sufficiently demonstrated their ability to multiply a whole number by 10, 100 or 1000.
- Occasionally ask the children to explain their method of working. Ask: **Louise, how did you get that answer?**

> Success criterion: *Divide a whole number by 10, 100 or 1000*

- Collect all the cards, reshuffle and once again evenly distribute the cards from RCMs 4–6 amongst the children.
- Say: **This time we're going to go around the group and when I call out 10, 100 or 1000 I want you to divide the number on your card by the number I call out. Ready?**
- Repeat the above, rotating around the group until the children have used all their cards or until each child has sufficiently demonstrated their ability to divide a whole number by 10, 100 or 1000.
- Occasionally ask the children to explain their thinking. Ask: **Jake, how did you work that out?**
- Place the cards from RCMs 4–6 to one side.

> Success criterion: *Multiply a decimal by 10, 100 or 1000*

● Evenly distribute the cards from RCMs 7 and 8 amongst the children.
● Say: **This time we're going to go around the group and I'm going to call out either 10, 100 or 1000. I then want you to multiply the number on your card by 10, 100 or 1000. Ready?**
● Repeat the above, rotating around the group until the children have used all their cards or until each child has sufficiently demonstrated their ability to multiply a decimal by 10, 100 or 1000.

> Success criterion: *Divide a decimal by 10, 100 or 1000*

● Collect all the cards, reshuffle and once again evenly distribute the cards from RCMs 7 and 8 amongst the children.
● Say: **This time when I call out either 10, 100 or 1000, I want you to divide the number on your card by 10, 100 or 1000. Ready?**
● Repeat the above, rotating around the group until the children have used all their cards or until each child has sufficiently demonstrated their ability to divide a decimal by 10, 100 or 1000.
● Once again occasionally ask the children to explain their thinking. Ask: **How do you know?**

● Why does 6×100 and 60×10 give the same answer?
● What number is 10 times as big as 1·05? What about 100 times bigger? …1000 times bigger?
● What number is 10 times smaller than 9? What about 100 times smaller? …1000 times smaller?
● When you multiply/divide a number by 10, 100 or 1000 in which direction do the digits move?

What to do for those children who achieve *above* expectation

● Place a selection of cards from RCMs 4–8 face up on the table in front of the children. Point to specific cards and ask the children to multiply and/or divide the numbers by 10, 100 or 1000, including decimals where the answer will result in four or more decimal places. Encourage the children to give quick responses (Level 5).

What to do for those children who achieve *below* expectation

● Avoid asking the children questions that will result in answers to three decimal places, e.g.
 – do not ask the children to divide a whole number by 1000,
 – only ask the children to multiply a decimal by 10 or 100,
 – when asking the children to divide a decimal, only use RCM 7 and divide by 10.

Task 15
Calculating
(Knowing and using number facts)

Objectives NC AT 2   NC Level 4

- Refine and use efficient written methods to multiply and divide HTU × U, TU × TU, U·t × U, and HTU ÷ U
- Use knowledge of rounding, place value, number facts and inverse operations to estimate and check calculations (optional)

Resources

- RCM 4: Two-digit number cards
- RCM 5: Three-digit number cards
- RCM 7: Decimal cards – tenths
- RCM 8: Decimal cards – hundredths (for children achieving above expectations)
- 0–9 die
- pencil and paper or RCM 16: Estimate, calculate and check (per child)

Task

NOTE: This task involves a number of Success criteria. It is advisable to choose just one or two criteria at a time.

- Prior to the task, shuffle and place each set of number cards into a pile of its own.
- Provide each child with a pencil and a piece of paper or RCM 16: Estimate, calculate and check. If you decide to use this RCM, ask the children to estimate and check each calculation.

Success criterion: *Refine and use efficient written methods to multiply HTU × U*

- Place a three-digit number card from RCM 5 in front of each child.
- Say: **I'm going to roll this die. I then want each of you to multiply the number on your card by the die number.**
- Allow the children sufficient time to complete their calculation, then ask: **What is the answer to your calculation? How did you get that answer?**
- Continue until each child has sufficiently demonstrated their ability to carry out an efficient written method to multiply a three-digit number by a one-digit number.

Success criterion: *Refine and use efficient written methods to multiply TU × TU*

- Place a pair of two-digit number cards from RCM 4 in front of each child.
- Say: **Using the pencil and paper I want each of you to multiply the two numbers together using a written method.**
- Allow the children sufficient time to complete their calculation, then ask: **Denis, what is your answer? How did you get your answer?**
- Continue until each child has sufficiently demonstrated their ability to carry out an efficient written method to multiply a pair of two-digit numbers.

Success criterion: *Refine and use efficient written methods to multiply U·t × U*

- Place a decimal card from RCM 7 in front of each child.
- Say: **I'm going to roll this die. I then want each of you to multiply the number on your card by the die number.**

- Allow the children sufficient time to complete their calculation, then ask: **What is the answer to your calculation? How did you get that answer?**
- Continue until each child has sufficiently demonstrated their ability to carry out an efficient written method to multiply a two-digit decimal with one place by a one-digit number.

> Success criterion: *Refine and use efficient written methods to divide HTU ÷ U*

- Place a three-digit number card from RCM 5 in front of each child.
- Say: **I'm going to roll this die. I then want each of you to divide the number on your card by the die number.**
- Allow the children sufficient time to complete their calculation, then ask: **What is the answer to your calculation? How did you get that answer?**
- Continue until each child has sufficiently demonstrated their ability to carry out an efficient written method to divide a three-digit number by a one-digit number.

- How did you work out the answer to the calculation? Which method did you use?
- How else could you have worked it out? Is there another way? Are there any other ways?
- What do you have to be aware of when multiplying numbers involving decimals?
- What is the most efficient way of working out the answer?
- Did you make an approximation of what your answer might be?
- Approximately what is the answer to the calculation?
- How could you check that your answer is correct?

What to do for those children who achieve *above* expectation

- Using the two-digit and three-digit number cards from RCMs 4 and 5, ask the children to multiply a three-digit number by a two-digit number.
- Using the hundredths decimal cards from RCM 8, ask the children to multiply a three-digit decimal with two decimal places by a one-digit number.

What to do for those children who achieve *below* expectation

- Using the two-digit number cards from RCM 4, ask the children to multiply and divide a two-digit number by a one-digit number.

Task 16
Calculating

Objective   NC AT 2   NC Level 4
- Find fractions using division, e.g. $\frac{1}{100}$ of 5 kg, and percentages of numbers and quantities, e.g. 10%, 5% and 15% of £80

Resources
- RCM 18: Finding fractions of numbers (enlarged to A3)
- RCM 19: Percentages 2 (enlarged to A3)
- pencil and paper clip (for the spinner on RCM 18)
- pencil and paper (per child)

Task

> Success criterion: *Find fractions using division*

- Place RCM 18 on the table in front of the children.
- Ask the children to take turns to spin the spinner, e.g. $\frac{1}{3}$.
- Looking at the fraction that the child has spun, e.g. $\frac{1}{3}$, point to a number on the RCM that is a multiple of the fraction's denominator, i.e. 6, 12, 15, 18, 21, …
- The child then uses the fraction as an operator to find the fraction of the number indicated, e.g. $\frac{1}{3} \times 15$.
- Occasionally ask the children to explain their method of working. Say: **Sophie, how did you work out that answer?**
- Continue until each child has sufficiently demonstrated their ability to find fractions using division.

> Success criterion: *Find percentages of numbers*

- Place RCM 19 on the table in front of the children.
- Point to one of the percentage labels at the top of the RCM, e.g. 25%, and one of the packages, e.g. 400.
- Ask a child to work out 25% of 400. Say: **Brian, I want you to work out what 25% of 400 is.**
- Repeat the above several times for each child.

NOTE: The first row of percentage labels is the easiest and the fourth row is the hardest.

- Occasionally ask the children to explain their method of working. Ask: **Sophie, how did you work out what 60% of 200 was?**
- Continue until each child has sufficiently demonstrated their ability to find simple percentages of numbers.

- How do you work out $\frac{1}{5}$ of 45?
- $\frac{1}{3}$ of a total is 8. What is the total? What other fractions of the total can you calculate?
- What percentages can you easily work out in your head? Why are they easy?
- When finding percentages of numbers or quantities what percentages do you use to help you? How do you use these percentages to work out others?
- Which percentages are easy/difficult to work out? Why?
- How would you find 20% of 80? Is there another way? Are there any others? Which do you find easier? Why?

What to do for those children who achieve *above* expectation

- After the children have found fractions and/or percentages of whole numbers, ask them questions that involve finding fractions and/or percentages of quantities, e.g.

  30% of £30

  $\frac{1}{3}$ of 20 kg

  30% of 4 m

  $\frac{1}{5}$ of 5 *l*

  40% of 10 cm

  $\frac{1}{6}$ of 1 hour

- When using RCM 19, choose percentages from the fourth row of percentage labels, i.e. 12%, 18%, 46%, 54% and 62%.

What to do for those children who achieve *below* expectation

- Do not spin the spinner on RCM 18, instead point to a simple unitary fraction, e.g. $\frac{1}{2}$ or $\frac{1}{5}$, and then to a number on the RCM, e.g. 12 or 15. Ask the children to use the fraction as an operator to find the fraction of the number indicated, i.e. $\frac{1}{2} \times 12$ or $\frac{1}{5} \times 15$.

- When using RCM 19, choose percentages from the first row of percentage labels, i.e. 1%, 10%, 25%, 50% and 75%.

Task 17
Calculating
(Knowing and using number facts)

Objectives
NC AT 2   NC Level 4

- Use a calculator to solve problems, including those involving decimals or fractions, e.g. to find $\frac{3}{4}$ of 150 g; interpret the display correctly in the context of measurement
- Use knowledge of rounding, place value, number facts and inverse operations to estimate and check calculations (optional)

Resources
- RCM 16: Estimate, calculate and check (per child)
- pencil (per child)
- calculator (per child)

Task

NOTE: Prior to the activity, write from four to six different calculations in the first column of RCM 16, similar to those below. The calculations should include a combination of addition, subtraction, multiplication and division, involving decimals or fractions and be appropriate to calculate using a calculator.

$2 \cdot 1$ m ÷ 6	3 kg ÷ 75
$12 \cdot 35 + 8 \cdot 2 + 17$	$3 \cdot 25 ÷ 5$
$\frac{3}{4} \times 150$	$\frac{2}{3} \times £12.60$
£8.56 × 5	□ × 84 = 4368
$12 \cdot 5 \times 8 \times 14 \cdot 6$	$53 \cdot 7 + 43 \cdot 8 - 82 \cdot 3$

> Success criteria: *Estimate results using a variety of strategies*
> *Use a calculator effectively*
> *Check results using a variety of strategies*

- Provide each child with a copy of RCM 16 containing from four to six different calculations, a pencil and a calculator.
- Ask the children to look at the calculations on their sheet.
- Tell the children that for each calculation you want them to:
  - estimate the answer first, writing their estimate and any working out in the second column
  - work out the answer using a calculator, writing down in the third column the keys pressed in order to get the answer in the third column
  - check their answer, writing any working out in the fourth column.

- As the children work through the calculations on the sheet ask them questions similar to those in the Assessment for Learning section, assessing their ability to use a calculator effectively.

- What keys would you press on a calculator to work out: 3 kg ÷ 75?
- What are you going to key into your calculator?
- Look at this calculation. What order are you going to key in the numbers and operations in this calculation?
- My calculator shows 4·5. What might the question have been? What if it was about money/length?

What to do for those children who achieve *above* expectation

- Ask the children to solve calculations involving percentages.

What to do for those children who achieve *below* expectation

- Use smaller numbers in the calculations to make estimating and mental approximations easier.

Answers

2·1 m ÷ 6 = 35 cm

12·35 + 8·2 + 17 = 37·55

$\frac{3}{4}$ × 150 = 112·5

£8.56 × 5 = £42.80

12·5 × 8 × 14·6 = 1460

3 kg ÷ 75 = 40 g

3·25 ÷ 5 = 0·65

$\frac{2}{3}$ × £12.60 = £8.40

52 × 84 = 4368

53·7 + 43·8 − 82·3 = 15·2

Task 18
Understanding shape

Objective   NC AT 3   NC Level 4
- Identify, visualise and describe properties of rectangles, triangles, regular polygons and 3-D solids; use knowledge of properties to draw 2-D shapes and identify and draw nets of 3-D shapes

Resources

- RCM 20: 2-D shapes and 3-D solids (enlarged to A3)
- RCM 21: Nets (enlarged to A3)
- set-square (per child)
- ruler (per child)
- pencil (per child)
- squared paper (per child)

Task

> Success criterion: *Identify, visualise and describe properties of rectangles, triangles, regular polygons and 3-D solids*

- Place RCM 20 in the middle of the table for all the children to see.
- Ask questions similar to those in the *Assessment for learning* section below that encourage children to identify, visualise and describe the properties of rectangles, triangles, regular polygons and 3-D solids. Children should be able to describe the following properties of 2-D shapes and 3-D solids:

2-D shapes
- Number of sides
- Number of angles
- Number of right angles
- Number of lines of symmetry

3-D solids
- Number of faces
- Number of edges
- Number of vertices (corners)

> Success criterion: *Use knowledge of properties to draw 2-D shapes*

- Provide each child with a set-square, ruler, pencil and a sheet of squared paper.
- Ask individual children to draw one of the following shapes:
  - pentagon
  - scalene triangle
  - square
  - equilateral triangle
  - hexagon

  - rectangle
  - isosceles triangle
  - octagon
  - right-angled triangle
  - heptagon
- If appropriate, ask each child to draw another 2-D shape.

> Success criterion: *Identify nets of 3-D solids*

- Place RCM 21 in the middle of the table for all the children to see.
- Ask: **What is a net?** (A flat shape which can be cut out and folded to make a 3-D solid.) Discuss the various responses offered by the children.
- Explain to the children that the first 24 shapes include nets of a closed cube, nets of an open cube and some do not make a cube at all. Say: **Some of these are nets of a closed cube and some are not.**

- Ask the children to identify which of the nets are of a closed cube. Say: **On the back of your sheet of squared paper, I want each of you to write down the numbers of those shapes that are the nets of a closed cube.**
- Allow the children sufficient time to complete this part of the task before telling them which of the shapes are nets of a closed cube.
- Repeat the above for the nets of a square-based pyramid.

- What is this shape called?
- Point to the octagon.
- Describe a tetrahedron to me.
- How many faces does a cuboid have? How many edges? ...vertices?
- Which of these triangles is an equilateral triangle? How do you know?
- How many lines of symmetry does an octagon have?

What to do for those children who achieve *above* expectation

- Ask the children to classify the shapes on RCM 20 according to their own, and given, criteria.

What to do for those children who achieve *below* expectation

- Only ask the children to identify, visualise and describe the properties of the following shapes:
  - 2-D shapes: square, rectangle, triangle, circle, pentagon, hexagon, heptagon and octagon
  - 3-D solids: cube, cuboid, sphere, cylinder, cone, pyramid
- Only ask the children to draw the following 2-D shapes: square, rectangle and triangle.

Answers
RCM 20: 2-D shapes and 3-D solids

a. cone
b. pentagon
c. scalene triangle
d. circle
e. cylinder
f. square
g. cube
h. equilateral triangle
i. cuboid
j. hexagon
k. sphere
l. rectangle
m. square-based pyramid
n. isosceles triangle
o. triangular-based pyramid (tetrahedron)
p. octagon
q. octahedron
r. right-angled triangle
s. heptagon

RCM 21: Nets
The following shapes are nets of a closed cube: 1, 7, 8, 10, 11, 15, 17, 20, 22, 23 and 24.
The following shapes are nets of a square-based pyramid: 1, 3, 4, 5, 6 and 8.

Task 19
Understanding shape

Objective NC AT 3 NC Level 4
* **Read and plot co-ordinates in the first quadrant; recognise parallel and perpendicular lines in grids and shapes; use a set-square and ruler to draw shapes with perpendicular or parallel sides**

Resources

- RCM 22: Co-ordinates and lines (per child)
- set of 3-D solid shapes
- pencil (per child)
- ruler (per child)
- red and blue coloured pencil (per child)
- set-square (per child)
- squared paper (per child)

Task

NOTE: This task involves a number of Success criteria. It is advisable to choose just one or two at a time.

> Success criterion: *Read and plot co-ordinates in the first quadrant*

- Provide each child with a copy of RCM 22, a pencil and a ruler.
- Referring to the co-ordinates grid at the top of the sheet, say: **You need to listen carefully. I am going to give you some instructions to follow. I will only say the instructions once, so you must listen carefully the first time. Do only the things you are told to do, and do nothing else. There are 8 instructions. Do each task immediately after the instructions have been given. Ready?**
- Say:
  1. **Draw a ring around the cross at co-ordinates (9, 3)**
  2. **Draw a ring around the cross at co-ordinates (2, 1)**
  3. **Draw a cross at the co-ordinates (6, 5)**
  4. **Draw a cross at the co-ordinates (3, 7)**
  5. **Draw a ring around the cross at co-ordinates (5, 2)**
  6. **Draw a cross at the co-ordinates (8, 1)**
  7. **Draw a ring around the cross at co-ordinates (8, 5)**
  8. **Draw a cross at the co-ordinates (4, 6)**

> Success criterion: *Recognise parallel sides in 2-D shapes*

- Ensure that each child has a copy of RCM 22, a pencil and a ruler. Also provide them with a red and blue coloured pencil.
- Referring to the 2-D shapes at the bottom of the sheet, ask the children to use their red pencil and a ruler to draw over the dotted lines in each shape to show a pair of parallel sides.
- Say: **Look at all the shapes on the sheet. I want you to look for a pair of parallel sides in each shape. Using your red pencil and a ruler I want you to draw over two lines that are parallel to each other. If you cannot find a pair of parallel sides in any of the shapes, then write a red zero inside that shape.**
- Next ask the children to identify how many pairs of parallel sides each shape has.
- Say: **Look at all the shapes. Now I want you to use your red pencil to write inside each shape how many pairs of parallel sides it has.**
- Continue until each child has sufficiently demonstrated their ability to recognise parallel sides in 2-D shapes.

Success criterion: *Recognise perpendicular sides in 2-D shapes*

● Now ask the children to use their blue pencil and a ruler to draw over the dotted lines in each shape to show a pair of perpendicular sides.
● Say: **Look at all the shapes on the sheet again. This time I want you to use the blue pencil and a ruler to draw over two lines on each shape that are perpendicular to each other. If you cannot find a pair of perpendicular sides in any of the shapes, then write a blue zero inside that shape.**
● Next ask the children to identify how many pairs of perpendicular sides each shape has.
● Say: **Look at all the shapes. Now I want you to use your blue pencil to write inside each shape how many pairs of perpendicular sides it has.**
● Continue until each child has sufficiently demonstrated their ability to recognise perpendicular sides in 2-D shapes.

Success criterion: *Recognise parallel and perpendicular edges in 3-D solids*

● Place the set of 3-D solids on the table in front of the children.
● Ask children questions similar to the following: **Look at this cube. How many edges are parallel to this one? How many edges are perpendicular to this one? How many vertices does a cuboid have? How many edges? How many pairs of parallel edges has a squared-based pyramid? …perpendicular edges…?**
● Continue until each child has sufficiently demonstrated their ability to recognise parallel and perpendicular edges in 3-D solids.

Success criterion: *Use a set-square and ruler to draw shapes with perpendicular or parallel sides*

● Ensure that each child has a pencil and a ruler. Also provide them with a set-square and a sheet of squared paper.
● Ask individual children to draw one of the following shapes: square, rectangle, hexagon, octagon, right-angled triangle or any polygon with a given number of perpendicular and/or parallel sides.
● If appropriate, ask each child to draw another 2-D shape.

● Which number comes first in a pair of co-ordinates, the x-axis or the y-axis?
● How did you find that point on the grid?
● How can you check to make sure two sides/edges are parallel/perpendicular?
● What do you know about the properties of a rectangle?
● What is the same/different about a square and a rectangle?
● Draw me a shape with one pair of parallel sides. Draw me a shape with two pairs of perpendicular sides.

What to do for those children who achieve *above* expectation

● Using their red and blue pencils, ask the children to mark all the pairs of parallel sides and perpendicular sides in each of the 2-D shapes.
● Ask the children to draw the following 2-D shapes: hexagon, octagon or any polygon with a given number of perpendicular and/or parallel sides.

What to do for those children who achieve *below* expectation

● Do not ask the children to identify all the pairs of parallel sides and perpendicular sides in each 2-D shape.

● Only ask the children to draw the following 2-D shapes: square, rectangle and triangle.

Answers
RCM 22: Co-ordinates and lines

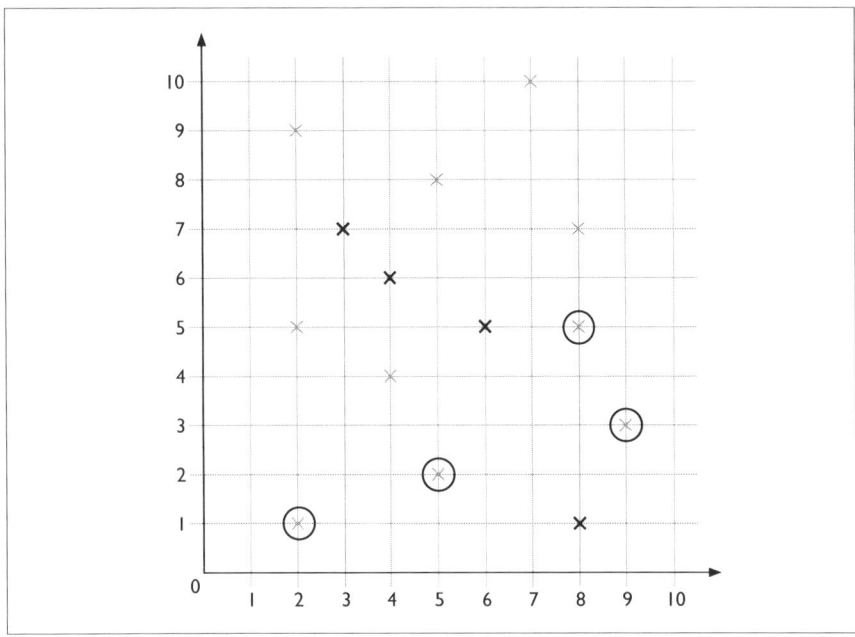

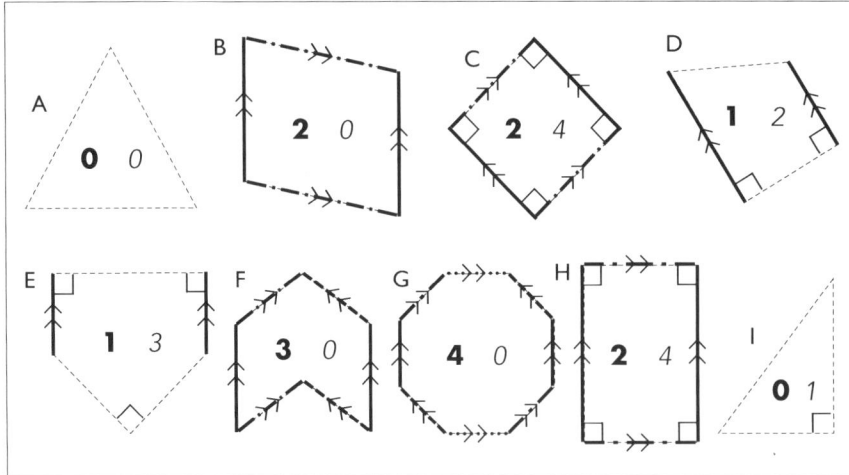

Bold number – Red answer (Number of pairs of parallel sides)

Italic number – Blue answer (Number of pairs of perpendicular sides)

Accept any of the marked pairs of parallel and perpendicular sides.

Task 20
Understanding shape

Objective   **NC AT 3**   **NC Level 4**
- Complete patterns with up to two lines of symmetry, and draw the position of a shape after a reflection or translation

Resources
- RCM 23: Symmetry, reflections and translations (per child)
- pencil (per child)
- ruler (per child)
- coloured pencil – optional (per child)

Task
- Provide each child with a copy of RCM 23, a pencil and a ruler.

> Success criterion: *Complete patterns with up to two lines of symmetry*

- Referring to Grid 1 on RCM 23, draw children's attention to the vertical and horizontal lines of symmetry.
- Ask the children to do one of the following:
  – Complete the pattern so that it is symmetrical in either the vertical or horizontal mirror lines – easier.
  – Complete the pattern so that it is symmetrical in both mirror lines – harder.

> Success criterion: *Draw the position of a shape after a translation*

- Referring to Grid 2 on RCM 23, ask the children to translate one of the following:
  – Shape A four squares to the right – easier.
  – Shape B five squares to the left and one square up – harder.
- If appropriate, ask the children to translate the other shape.

> Success criterion: *Draw the position of a shape after a reflection*

- Referring to Grid 3 on RCM 23, ask the children to reflect one of the following:
  – Shape C along the horizontal line of symmetry – easier.
  – Shape D along the horizontal line of symmetry – harder.
- If appropriate, ask the children to reflect the other shape.

- Explain to me how this shape has been reflected. …translated.

What to do for those children who achieve *above* expectation
- Ask the children to complete the three harder reflections and translations.

What to do for those children who achieve *below* expectation
- Ask the children to complete the three easier reflections and translations.

Answers

RCM 23: Symmetry, reflections and translations

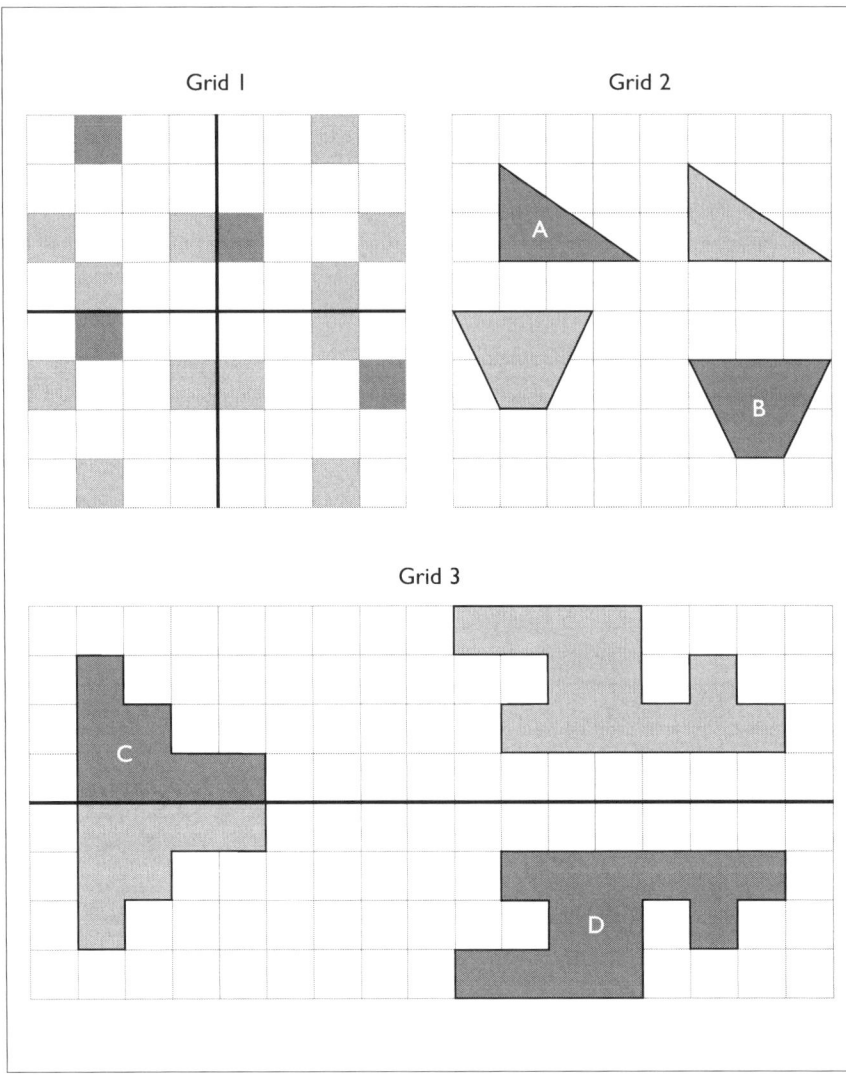

Grid 1

Grid 2

Grid 3

Task 21
Understanding shape

Objective	NC AT 3	NC Level 4

- Estimate, draw and measure acute and obtuse angles using an angle measurer or protractor to a suitable degree of accuracy; calculate angles in a straight line

Resources

- RCM 24: Measuring angle cards (enlarged to A3 and cut out)
- RCM 25: Angles in a straight line cards (enlarged to A3 and cut out)
- pencil and paper (per child)
- ruler (per child)
- protractor (per child)

Task

- Provide each child with a pencil and a piece of paper, a ruler and a protractor.

> Success criterion: *Estimate acute and obtuse angles*

- Place one of the cards from RCM 24 in front of each child. See below for guidance as to which card to give to individual children depending on their ability.

	Easy (angles to the nearest 10°)	**Moderate** (angles to the nearest 5°)	**Difficult** (angles to the nearest 1°)
Cards	1, 2, 3, 4	5, 6, 7, 8	9, 10, 11, 12

- Ask: **I want each of you to look at the angle drawn on the card in front of you and to tell me whether it is an acute, obtuse or right-angle.**
- Once each child has done this, say: **Now I want each of you to estimate the size of the angle on the card.**

> Success criterion: *Measure acute and obtuse angles using a protractor*

- Now ask each child to use their protractor to measure the size of the angle.
- Repeat the above until each child has sufficiently demonstrated their ability to estimate and measure angles.

> Success criterion: *Draw acute and obtuse angles using a protractor*

- Ask individual children to use their ruler and protractor to draw one of the following angles. See below for guidance as to which angle to give to individual children depending on their ability.

Easy (angles to the nearest 10°)	**Moderate** (angles to the nearest 5°)	**Difficult** (angles to the nearest 1°)
20°, 70°, 90°, 100°	35°, 65°, 95°, 115°	12°, 28°, 76°, 121°

- Continue until each child has sufficiently demonstrated their ability to draw angles.

Success criterion: *Calculate angles in a straight line*

● Using the cards from RCM 25, place one card in front of each child. See below for guidance as to which card to give to individual children depending on their ability.

	Easy (angles to the nearest 10°)	**Moderate** (angles to the nearest 5°)	**Difficult** (angles to the nearest 1°)
Cards	1, 2, 3	4, 5, 6, 7, 8, 9	10, 11, 12

● Say: **Look at the card in front of you. I want each of you to calculate the size of the missing angle.**
● Once the children have done this, ask: **How did you calculate the size of the missing angle? How did you know it was ... degrees?**

● What things do you have to remember when you use a protractor to measure or draw angles?
● What is the angle between the hands of a clock at nine o'clock? Explain how you know.
● Estimate the size of each of these angles. Now measure them to the nearest degree. How close were you?
● Draw me an angle of 55°.
● How many degrees are there in a straight line?

What to do for those children who achieve *above* expectation

● Give the children cards 9–12 from RCM 24 to measure angles to the nearest 1°.
● Ask the children to draw angles to the nearest 1°.
● Give the children cards 10, 11 and 12 from RCM 25 to calculate angles in a straight line.

What to do for those children who achieve *below* expectation

● Give the children cards 1–4 from RCM 24 to measure angles to the nearest 10°.
● Give the children cards 5–8 from RCM 24 to measure angles to the nearest 5°.
● Ask the children to draw angles to the nearest 10° or 5°.
● Give the children cards 1, 2 and 3 from RCM 25 to calculate angles in a straight line.

Answers

RCM 24: Measuring angle cards

1. 50°	**2.** 60°	**3.** 130°
4. 150°	**5.** 85°	**6.** 25°
7. 105°	**8.** 155°	**9.** 53°
10. 38°	**11.** 82°	**12.** 147°

RCM 25: Angles in a straight line cards

1. $a = 120°$	**2.** $b = 70°$	**3.** $c = 130°$
4. $d = 145°$	**5.** $e = 55°$	**6.** $f = 105°$
7. $g = 75°$	**8.** $h = 65°$	**9.** $i = 60°$
10. $j = 107°$	**11.** $k = 61°$	**12.** $l = 64°$

Task 22
Measuring

Objective   **NC AT 3**   **NC Level 4**

• Read, choose, use and record standard metric units to estimate and measure length, weight and capacity to a suitable degree of accuracy, e.g. the nearest centimetre; convert larger to smaller units using decimals to one place, e.g. change 2.6 kg to 2600 g

Resources

- *Length:*
  - about 6 objects measuring between 10 cm and 2 m
  - ruler
  - tape measure
  - metre stick
- *Weight:*
  - about 6 objects weighing between 1 kg and 5 kg
  - set of scales
- *Capacity:*
  - about 6 empty containers of different shapes and sizes between 250 ml and 5 litres
  - bucket of water
  - 2 litre measuring jug
  - funnel
- 3 large sheets of paper, labelled as follows:

Length				
Object	Instrument	Unit	Estimated length	Actual length

Weight				
Object	Instrument	Unit	Estimated weight	Actual weight

Capacity				
Object	Instrument	Unit	Estimated capacity	Actual capacity

- marker
- RCM 26: Converting measures cards (enlarged to A3 and cut out)
- pencil and paper (per child)

Task

NOTES: This task is best undertaken with only two or three children.

This task involves a number of Success criteria. It is advisable to choose just one or two at a time.

> Success criterion: *Read, choose, use and record standard metric units to estimate and measure length to a suitable degree of accuracy*

- Show the children the objects measuring between 10 cm and 2 m, ruler, tape measure, metre stick and the large sheet of paper headed: 'Length'.
- Referring to specific objects, ask children questions similar to the following:
  - **How long do you think the … is?**
  - **How could you find out how long it is?**
  - **What could you use to measure it?**
  - **What units of measure would be suitable / would not be suitable for measuring it?**
- As children answer the questions, ask them to record the information in the table on the large sheet of paper.
- Ask individual children to find the length of specific objects. Say: **Louise, measure the length of the …**
- Continue asking each child to measure the length of one or more objects and to record the information on the chart.

> Success criterion: *Read, choose, use and record standard metric units to estimate and measure weight to a suitable degree of accuracy*

- Show the children the objects weighing between 1 kg and 5 kg, the set of scales and the large sheet of paper headed: 'Weight'.
- Referring to specific objects, ask children questions similar to the following:
  - **What do you think is the weight of the …?**
  - **How could you find out how heavy it is?**
  - **What could you use to weigh it?**
  - **What units of measure would be suitable / would not be suitable for weighing it?**
- As children answer the questions, ask them to record the information in the table on the large sheet of paper.
- Ask individual children to find the weight of specific objects. Ask: **Louise, what is the weight of the …?**
- Continue asking each child to find the weight of one or more objects and to record the information on the chart.

> Success criterion: *Read, choose, use and record standard metric units to estimate and measure capacity to a suitable degree of accuracy*

- Show the children the empty containers of different shapes and sizes between 250 ml and 5 *l*, the bucket of water, 2 litre measuring jug, funnel and the large sheet of paper headed: 'Capacity'.
- Referring to specific containers, ask children questions similar to the following:
  - **How much do you think this container can hold?**
  - **How could you find out how much it can hold?**
  - **What could you use to find out?**
  - **What units of measure would be suitable / would not be suitable for finding out how much it can hold?**
- As children answer the questions, ask them to record the information in the table on the large sheet of paper.

- Ask individual children to find the capacity of specific containers. Ask: **Jasmine, can you find out for me how much this container can hold?**
- Continue asking each child to find the capacity of one or more containers and to record the information on the chart.

> Success criterion: *Convert larger to smaller units, and smaller to larger units, using decimals to one place*

- Shuffle the cards from RCM 26 and place them face down in a pile. Provide each child with a pencil and a piece of paper.
- Turn over the top card, e.g. 5·2 kg, and ask the children to convert the measurement into a smaller or larger unit. Ask: **What is 5 point 2 kilograms in grams? Each of you, write this for me on your piece of paper.**
- Repeat for the other cards, e.g. 720 cm, ask: **Write down what 720 centimetres is in metres. What about in metres and centimetres?**
- Continue until each child has sufficiently demonstrated their ability to convert larger to smaller units (and vice versa).

- What do you think is the length/weight/capacity of the …?
- How could you find out how long/heavy it is? … how much water it can hold? What could you use?
- What units of measure would be suitable / would not be suitable?
- How do I write 5 metres 8 centimetres as a decimal?
- Tell me an example of something you would measure in kilometres. What about metres / centimetres / millimetres?
- How do I write 14 kilograms 600 grams as a decimal? What about 8 kilograms and 20 grams?
- Tell me an example of something you would measure in kilograms/grams. What about grams? What about litres/millilitres?

What to do for those children who achieve *above* expectation

- Encourage the children to be as accurate as possible when estimating and measuring different lengths, weights and capacities.
- Ask children to convert larger to smaller units, and smaller to larger units, using decimals to two places.

What to do for those children who achieve *below* expectation

- Only ask the children to convert larger to smaller units, and smaller to larger units, using whole numbers.

Task 23

Measuring

Objective   NC AT 3   NC Level 4
- Interpret a reading that lies between two unnumbered divisions on a scale

Resources
- several copies of RCM 27: Scales (enlarged to A3)
- pencil

Task

NOTES:

- Prior to the task, mark different lengths, masses and capacities on the various scales on RCM 27, e.g.

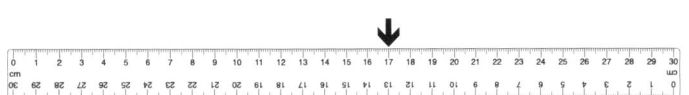 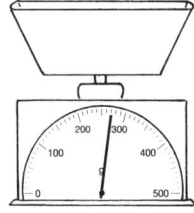 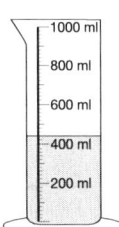

- You may wish to mark up more than one copy of RCM 27 for this task. You will also need several blank copies of RCM 27.
- Vary the markings on the scales according to the ability of the children undertaking in the task, i.e.
  - mark a numbered division – easy
  - mark an unnumbered division – moderate
  - mark between two unnumbered divisions – difficult.

> Success criterion: *Interpret and record a reading that lies between two unnumbered divisions on a scale – Length*

- Referring to the different rulers on RCM 27, ask questions that require the children to interpret the different readings. Point and say: **What length is the arrow pointing to on this ruler?**
- Mark a second point on the ruler and ask: **What is the distance between these two arrows?**
- Referring to a blank copy of RCM 27, ask individual children to record a length on one of the rulers. Point and say: **Leroy, show me where 65 cm is on this ruler.**
- Continue until each child has sufficiently demonstrated their ability to interpret and record readings on scales involving length.

> Success criterion: *Interpret and record a reading that lies between two unnumbered divisions on a scale – Mass*

- Referring to the different weighing scales on RCM 27, ask questions that require the children to interpret the different readings. Point and say: **What weight is the arrow pointing to?**
- Referring to a blank copy of RCM 27, ask individual children to record a weight on one of the scales. Point and say: **Natasha, if something weighed 1.4 kg, what would this look like on the scales?**
- Continue until each child has sufficiently demonstrated their ability to interpret and record readings on scales involving weight.

Success criterion: *Interpret and record a reading that lies between two unnumbered divisions on a scale – Capacity*

- Referring to the different containers on RCM 27, ask questions that require the children to interpret the different readings. Point and say: **What is the water level in this container?**
- Referring to a blank copy of RCM 27, ask individual children to record a capacity on one of the containers. Point and say: **Lance, if this container had 240 ml of liquid in it what would this look like?**
- Continue until each child has sufficiently demonstrated their ability to interpret and record readings on scales involving capacity.

- What length is this ruler showing?
- What is the distance between these two points?
- What weight is this set of scales showing?
- Show me a weight of 340 g.
- How much liquid is in this container?
- Show me what this container would look like if it had 620 ml of liquid in it.

What to do for those children who achieve *above* expectation

- Vary the markings on the scales according to the ability of the children undertaking the task, i.e. mark between two unlabelled divisions.

What to do for those children who achieve *below* expectation

- Vary the markings on the scales according to the ability of the children undertaking the task, i.e. mark a labelled division.

Task 24
Measuring

Objective   NC AT 3   NC Level 4

- **Draw and measure lines to the nearest millimetre; measure and calculate the perimeter of regular and irregular polygons; use the formula for the area of a rectangle to calculate the rectangle's area**

Resources

- RCM 28: Polygon cards
- RCM 29: Rectangle cards
- large sheet of paper and marker
- ruler (per child)
- pencil and paper (per child)

Task

- Prior to the task, on the large sheet of paper, use a ruler to draw different lines of various lengths and orientation, i.e. vertical, horizontal and diagonal, to the nearest millimetre.

> Success criterion: *Draw lines to the nearest millimetre*

- Provide each child with a ruler, pencil and a piece of paper.
- Ask individual children to draw lines of various lengths. Say: **Michael, on your sheet of paper, I want you to draw a line 12.6 centimetres long. Fabio, I want you to draw a line 224 millimetres long.**
- Repeat the above until each child has sufficiently demonstrated their ability to draw lines to the nearest millimetre.

> Success criterion: *Measure lines to the nearest millimetre*

- Show the children the large sheet of paper showing different lines of various lengths and orientation.
- Ask individual children to measure the length of different lines. Say: **Michael, measure the length of this line.**
- Repeat the above until each child has sufficiently demonstrated their ability to measure lines to the nearest millimetre.

> Success criterion: *Measure and calculate the perimeter of regular and irregular polygons*

- Ask: **What does the word 'perimeter' mean?** (The distance all the way round the edge of something: the boundary.) Discuss the various responses offered by the children.
- Place one of the cards from RCM 28 in front of each child. See below for guidance as to which card to give to individual children depending on their ability.

	Easy (3- or 4-sided polygon)	**Moderate** (4- or 5-sided polygon)	**Difficult** (5-, 6- or 8-sided polygon)
Cards	1, 2, 3	4, 5, 6	7, 8, 9

- Say: **I want each of you to use your ruler to measure the length of each side of your shape and to then calculate the shape's perimeter.**
- If necessary, repeat the above until each child has sufficiently demonstrated their ability to measure and calculate the perimeter of regular and irregular polygons.

Success criterion: *Use the formula for the area of a rectangle to calculate its area*

- Ask: **What does the word 'area' mean?** (The amount of surface space inside the perimeter.) Discuss the various responses offered by the children.
- Place one of the cards from RCM 29 in front of each child. See below for guidance as to which card to give to individual children depending on their ability.

	Easy (U × U rectangle)	**Moderate** (U·5 × U rectangle)	**Difficult** (U·t × U rectangle)
Cards	1, 2, 3	4, 5, 6	7, 8, 9

- Say: **I want each of you to use your ruler to measure the length of each side of your rectangle and to then calculate the shape's area.**
- If necessary, repeat the above until each child has sufficiently demonstrated their ability to measure and calculate the area of a rectangle.

- Draw each of these lines using a 30 cm ruler: 8·3 cm, 0·9 cm, 46 mm, 8 mm.
- Measure the sides of these polygons in centimetres and millimetres. What is the perimeter of each shape in centimetres? …in millimetres?
- What is the perimeter of:
  – a regular octagon with sides of 25 mm? How did you know?
  – an equilateral triangle with sides of 6·4 cm? How did you work it out?
- A square has a perimeter of 56 cm. How long is each side? How did you work it out?
- A rectangle has a perimeter of 38 m. The shortest side is 7 m long. What is the length of the longest side? How did you know?

What to do for those children who achieve *above* expectation

- Give the children cards 7, 8 or 9 from RCM 28 to measure and calculate the perimeter of regular and irregular polygons.
- Give the children cards 7, 8 or 9 from RCM 29 to calculate the area of a rectangle.

What to do for those children who achieve *below* expectation

- Ask children to draw lines to the nearest centimetre.
- Give the children cards 1, 2 or 3 from RCM 28 to measure and calculate the perimeter of regular and irregular polygons.
- Give the children cards 1, 2 or 3 from RCM 29 to calculate the area of a rectangle.

Answers

RCM 28: Polygon cards

1. 15 cm   **2.** 16·5 cm   **3.** 14 cm
4. 12·8 cm   **5.** 18 cm   **6.** 18 cm
7. 16 cm   **8.** 16·4 cm   **9.** 15·5 cm

RCM 29: Rectangle cards

1. 5 cm × 4 cm = 20 cm²   **2.** 4 cm × 4 cm = 16 cm²   **3.** 6 cm × 4 cm = 24 cm²
4. 6.5 cm × 3 cm = 19·5 cm²   **5.** 4.5 cm × 3 cm = 13·5 cm²   **6.** 4 cm × 5·5 cm = 22 cm²
7. 6.8 cm × 2 cm = 13·6 cm²   **8.** 5.7 cm × 4 cm = 22·8 cm²   **9.** 2.4 cm × 6 cm = 14.4 cm²

Task 25
Measuring

Objective   NC AT 3   NC Level 4
- Read timetables and time using 24-hour clock notation; use a calendar to calculate time intervals

---

Resources
- demonstration 24-hour digital clock
- RCM 30: Timetable and calendar (enlarged to A3)

Task

> Success criterion: *Read the time using 24-hour clock notation*

- Show a time on the 24-hour digital clock, e.g. 04:37. Ask: **Toni, using a.m. and p.m. notation, can you tell me what time this clock shows?** (4:37 a.m.) **How else could you say this time?** (23 minutes to 5 in the morning).
- Repeat several times until each child has sufficiently demonstrated their ability to read the time using 24-hour clock notation.

> Success criterion: *Display the time using 24-hour clock notation*

- Ask: **Thomas, what would quarter past five in the afternoon look like on a 24-hour digital clock?** (17:15)
- Repeat several times until each child has sufficiently demonstrated their ability to display the time using 24-hour clock notation.

> Success criterion: *Read and interpret timetables*

- Show the children the timetable for London to Penzance from RCM 30.
- Ask questions similar to the following that require the children to read and interpret the timetable.
  - **If I caught the 12:05 train from London, at what time would I arrive at Plymouth? ...Bodmin? ...Penzance?**
  - **If I catch the 13:05 from London to Penzance, how many stops are there?**
  - **How long does it take to travel from Reading to Exeter if I catch the 14:05 train?**
  - **How long does the 07:30 train from London to Penzance take?**
  - **Approximately how long does it take to travel from St. Erth to Penzance?**
  - **If I lived in London and needed to be in Redruth for 7 o'clock in the evening, what is the latest train I could catch?**
- Ask the children to make statements about the timetable. Say: **Look at this timetable; tell me anything you notice about it. Two things that I notice are that there are ten trains a day from London to Penzance but only eight of them stop at Newton Abbot.**
- Continue asking questions until each child has sufficiently demonstrated their ability to read and interpret timetables.

Success criterion: *Read and interpret a calendar*

- Show the children the calendar from RCM 30.
- Ask questions similar to the following that require the children to read and interpret the calendar, and calculate time intervals.
  - **What day of the week is the 6th of May?**
  - **What date is the second Monday in October?**
  - **How many Friday's are there in February?**
  - **How many days are there in June?**
  - **In which month is the 24th a Tuesday?**
  - **How many days are there from March 25th until May 5th?**
  - **It is Peter's birthday on 29th September. If the date today was August 10th, how many more days until Peter's birthday?**
- Ask the children to make statements about the calendar. Say: **Look at this calendar; tell me anything you notice about it. One thing that I notice is that my birthday is the 14th of November, and on this calendar that falls on a Sunday.**
- Continue asking questions until each child has demonstrated their ability to read and interpret a calendar, and calculate time intervals.

- How long does it take to travel from Bodmin to St. Erth on the 15:05 train?
- How many stops are there between Exeter and Redruth?
- At what station is the train at 20:36?
- On what day of the week is the 18th of June?
- How many Wednesdays are there in August?
- How many days are there from the 3rd of August until the 20th of October?

What to do for those children who achieve *above* expectation

- Ask questions that involve reading and interpreting the timetable and calendar that are appropriate to the children's ability.

What to do for those children who achieve *below* expectation

- Ask questions that involve reading and interpreting the timetable and calendar that are appropriate to the children's ability.

Task 26
Handling data

Objective   NC AT 4   NC Level 4
• Describe the occurrence of familiar events using the language of chance or likelihood

Resources
- RCM 31: Probability (enlarged to A3)
- 1 set of 0–9 digit cards

Task

Success criterion: *Describe the occurrence of familiar events using the language of chance or likelihood*

- Show the children RCM 31. Remind the children of the five different probabilities on the sheet.
- Pointing to each of the five different probabilities in turn, ask: **Simone, can you tell me something that is certain to happen? Toby, can you tell me something that is likely to happen?**
- Continue to ask children questions that require them to suggest an event for each of the five probabilities on the sheet.
- Then ask children questions that require them to describe the chance or likelihood of an event, e.g. **What is the probability that we will do PE today? What is the probability that it will rain today? What is the probability that the number rolled on a 1–6 die will be an even number?**
- Next, show the children the set of 0–9 digit cards. Shuffle the cards and lay them out in a line face down in the middle of the table.
- Say: **We are going to play a game. I'm going to turn over the first card. We are then going to discuss the probability of the next card in the line being higher or lower. Each of you are then going to have to decide whether you think the card will be lower or higher, and then I'll turn over the next card. If you are correct you score a point. We'll keep going like this until all 10 cards have been turned over and we'll see who has scored most points.**
- Play the game with the children. As the children play the game, assess individual children's ability to use the language of probability.

- What is the probability of tossing a coin and it landing tails up?
- Tell me something that is likely to happen today?

What to do for those children who achieve *above* expectation

● On RCM 31, make the probabilities into a probability scale, i.e.

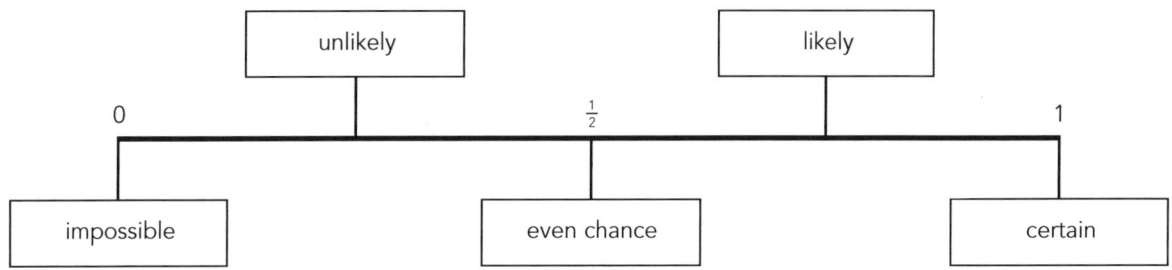

Ask the children to describe and predict outcomes from data using the language of chance or likelihood and the probability scale 0 to 1 (Level 5).

What to do for those children who achieve *below* expectation

● Only ask the children to describe the occurrence of familiar events using the language of certain, even chance and impossible.

Task 27
Handling data
Using and applying mathematics

Objectives   NC AT 4   NC Level 4

- Answer a set of related questions by collecting, selecting and organising relevant data; draw conclusions, using ICT to present features, and identify further questions to ask
- **Construct frequency tables, pictograms and bar and line graphs to represent the frequencies of events and changes over time**
- Plan and pursue an enquiry; present evidence by collecting, organising and interpreting information; suggest extensions to the enquiry

Resources

- RCM 32: Collecting, selecting and organising data (per child)
- squared and/or graph paper (per child)
- pencil (per child)
- ruler (per child)
- ICT data handling package – optional (per child or group)

Task

NOTE: You may want the children to work in pairs for this task.

- Prior to the task, write a question in the box at the top of RCM 32. Choose a topic to investigate that is relevant to your particular circumstances and of interest to the children, e.g. *What is our favourite food? What is the average height of a 10 year old? What after-school activities are the most popular? How far do most children in Year 5 live from the school? What are the five most favourite films in our school? What is our favourite month of the year? How many brothers and sisters do most children in our school have?* Alternatively, you may wish the children to suggest their own line of enquiry.

  Success criteria: *Collect data*
  Select data
  Organise data
  Construct frequency tables, pictograms and bar and line graphs
  Interpret data
  Draw conclusions
  Identify further questions to ask

- Provide each child with a copy of RCM 32, some squared and/or graph paper, a pencil and a ruler.
- Briefly discuss the question with the children.
- Discuss with the children:
  – their initial ideas of what to do
  – how they are going to collect the data
  – how they are going to organise the data.
- Ask the children to collect, organise and present the data.
- When the children have done this, they write about what they found out and what else they could find out about the topic.

- What information will you need to collect to answer these questions? How will you collect it? How will you display your data?
- What does this graph tell you? What makes the information easy or difficult to interpret?
- Make up three questions that can be answered using the data in this table/graph/chart.
- What further information could you collect to answer the question more fully?
- Which tables/charts/graphs are easy/difficult to interpret information from? Why?
- How did you work out that answer? Which information in the table/chart/graph did you use?

What to do for those children who achieve *above* expectation
- Ask the children to present their data using an ICT data handling package.

What to do for those children who achieve *below* expectation
- Ask the children to work in pairs.

Task 28
Handling data

Objective   **NC AT 4**   **NC Level 4**
• Find and interpret the mode of a set of data

Resources
- RCM 33: Mode (enlarged to A3)
- 1–6 die, 1–10 die, 1–12 die, 1–20 die

Task

> Success criterion: *Find and interpret the mode of a set of data*

- Place RCM 33 in the middle of the table. Referring to the first row of numbers at the top of the sheet, say: **Look at these numbers, what is the mode?** (3) **How do you know?**
- Remind the children that the term 'mode' refers to the most common or popular value.
- Repeat for the other 3 rows of numbers. (2nd row: 6, 3rd row: 5, 4th row: 7 and 3)
- Roll a 1–6 die. With the children's help record the results in the appropriate table.
- Repeat about 10 times. Ask: **What is the mode?**

- Continue to roll the die several times more, and record the results in the frequency table. Ask: **What is the mode now? Has it changed?** Complete the frequency table.
- Repeat the above using the 1–10 die and record the results in the appropriate table.
- Repeat the above using the 1–12 die and record the results in the appropriate table.
- Repeat the above using the 1–20 die and record the results in the appropriate table.

- How do you find the mode of a set of data?
- What does the mode tell you about a set of data?

What to do for those children who achieve *above* expectation
- Using the data presented in Task 27, where appropriate, ask the children to work out the mode.
- Ask the children to calculate the range (Level 4), median and mean (Level 5)

What to do for those children who achieve *below* expectation
- Roll the dice fewer times and ask the children to say the mode.

Self assessment Unit A1

Name _____ Date _____

- I can find missing numbers in a sequence that includes negative numbers ☺ 😐 ☹
- I can order a given set of whole numbers ☺ 😐 ☹
- I can partition, round and order decimals with one place ☺ 😐 ☹
- I can say what any digit represents in a number with up to seven digits ☺ 😐 ☹
- I can work out sums and differences of pairs of two-digit numbers ☺ 😐 ☹
- I can explain each step when I write addition and subtraction calculations in columns ☺ 😐 ☹
- I know my tables to 10. I can use them to work out division facts ☺ 😐 ☹
- I can find a pair of factors for a two-digit number ☺ 😐 ☹
- I can multiply or divide a whole number by 10, 100 or 1000 ☺ 😐 ☹
- I can work out two-digit × one-digit calculations in my head or with jottings; I can explain how I found the answer ☺ 😐 ☹
- I can estimate and check the result of a calculation ☺ 😐 ☹

Self assessment Unit B1

Name _____ Date _____

- I can make estimates when adding and subtracting large numbers ☺ 😐 ☹

- I can use written methods to add and subtract whole numbers and decimals, including money ☺ 😐 ☹

- I can explain each step when I write addition and subtraction calculations in columns ☺ 😐 ☹

- I can check whether a calculation is correct and explain how I did this ☺ 😐 ☹

- I know my tables to 10. I can use them to work out division facts ☺ 😐 ☹

- I can use tables facts to multiply multiples of 10 and 100 ☺ 😐 ☹

- I can find pairs of factors that multiply to make a given number ☺ 😐 ☹

- I can find a number that is a multiple of two different numbers ☺ 😐 ☹

- I can sort shapes according to their properties and explain how I sorted them ☺ 😐 ☹

- I know the important features of a cube and can use these to draw its net ☺ 😐 ☹

- I can describe the important features of shapes such as rectangles ☺ 😐 ☹

Self assessment Unit C1

Name _____ Date _____

- I can measure lengths using appropriate measuring instruments
- I can state measurements in km, m, cm and mm
- I can convert from one unit of length to another
- I can use decimals to record lengths
- I can measure and draw lines in millimetres
- I can collect and organise data to find out about a subject or to answer a question
- I can present and interpret information in a table
- I can present and interpret information in a pictogram
- I can present and interpret information in a bar chart
- I can present and interpret information in a bar line chart
- I can explain why I chose to represent the data using a particular table, graph or chart
- I know that the 'mode' is the most common piece of information
- I can find the mode of a set of data that I have collected

Collins New Primary Maths

Self assessment Unit D1

Name _____ Date _____

- I can identify the steps I need to take to solve problems ☺ ☺ ☹

- I can decide whether to do a calculation using mental methods, written methods or a calculator ☺ ☺ ☹

- I can use a calculator to solve problems that involve decimal measurements ☺ ☺ ☹

- I can multiply and divide whole numbers by 10, 100 and 1000 ☺ ☺ ☹

- I can choose appropriate units to measure length and distance ☺ ☺ ☹

- I can read metre sticks, tape measures and rulers marked in cm and mm accurately ☺ ☺ ☹

- I know how many millimetres there are in a centimetre or metre, and how many metres there are in a kilometre ☺ ☺ ☹

- I can make sensible estimates of length in everyday contexts ☺ ☺ ☹

- I can interpret a reading between two unnumbered divisions on a ruler, tape measure or metre stick ☺ ☺ ☹

- I can draw and measure lines to the nearest millimetre ☺ ☺ ☹

- I can measure the sides of polygons and find the perimeter ☺ ☺ ☹

- I can change a.m. or p.m. times to 24-hour clock times, and vice versa ☺ ☺ ☹

- I can use a calendar ☺ ☺ ☹

- I can read and plot co-ordinates ☺ ☺ ☹

Collins New Primary Maths

Self assessment Unit E1

Name _____ Date _____

- I can break a problem into steps and say the calculation I need to do to work out each step   ☺ 😐 ☹

- I check that my answer is sensible   ☺ 😐 ☹

- I can decide whether to solve problems using mental, written or calculator methods and explain my choice   ☺ 😐 ☹

- I can use diagrams to check that two fractions are equivalent   ☺ 😐 ☹

- I can explain how I know that two fractions, such as $\frac{7}{10}$ and $\frac{14}{20}$, are equivalent   ☺ 😐 ☹

- I can recognise fractions and decimals that are the same   ☺ 😐 ☹

- I can find fractions of numbers using division. For example, to find $\frac{1}{3}$ of a number, I divide it by 3   ☺ 😐 ☹

- I know what calculations to enter into a calculator to find a fraction of an amount   ☺ 😐 ☹

- I can find percentages of numbers   ☺ 😐 ☹

- I can use multiplication and division facts to multiply and divide multiples of 10 and 100   ☺ 😐 ☹

- I can find pairs of factors that multiply to make a given number   ☺ 😐 ☹

- I use different mental strategies for multiplication and division depending on the numbers involved. I can explain why I chose a particular method   ☺ 😐 ☹

- I can solve multiplication calculations using written methods. I can explain each step   ☺ 😐 ☹

Collins New Primary Maths

Self assessment Unit A2

Name _____ Date _____

● I can explain clearly to others my method for solving a problem. I listen to other children's methods. I talk about which is the most efficient method

● I can say what any digit in a decimal is worth

● I can explain why I chose to work mentally, use written methods or a calculator

● I can work out sums and differences of whole numbers or decimals

● I can explain each step when I add or subtract whole numbers or decimals using a written method

● I can double and halve decimals with two digits

● I know my tables to 10 for multiplication facts and division facts

● I can use multiplication and division facts to divide multiples of 10 and 100

● I can multiply or divide numbers by 10, 100 or 1000

● I can apply mental strategies to multiply by 12, 19 and 21

● I can use a calculator to solve a problem, I can explain what calculations I keyed into the calculator and why

● I can estimate and check the result of a calculation

Self assessment Unit B2

Name _____ Date _____

● I can split a word problem into steps and work out what calculation to do for each step. I can explain what the answer to each step tells me

● I can add or subtract whole numbers or decimals in my head by using a related two-digit addition or subtraction

● I can investigate a general statement and say whether examples are true or false

● I can use tables facts and linked division facts

● I can double and halve whole numbers

● I can find the double or half of a decimal by doubling or halving the related whole number

● I can say whether a triangle is equilateral, isosceles or scalene and explain how I know

● I can recognise parallel and perpendicular lines

● I can explain whether a shape has line symmetry

● I can reflect a shape through different lines of symmetry

Collins
New
Primary
Maths

Self assessment Unit C2

Name _____ Date _____

● I understand weights presented as whole numbers, fractions or decimals that are the same   ☺ ☺ ☹

● I can find the value of each interval on a scale and use this to give approximate values of readings between divisions   ☺ ☺ ☹

● I collect and organise data to find out about a subject or to answer a question   ☺ ☺ ☹

● I use graphs to show findings about a subject or to help explain my answer to a question   ☺ ☺ ☹

● I can decide what information needs to be collected to answer a question and how best to collect it   ☺ ☺ ☹

● I can explain what a table, graph or chart tells us and consider questions that it raises   ☺ ☺ ☹

● I can explain why I chose to represent data using a particular table, graph or chart   ☺ ☺ ☹

● I can present and interpret data in a bar line graph   ☺ ☺ ☹

● I can present and interpret data in a line graph   ☺ ☺ ☹

● I can describe how likely an event is to happen and justify my statement   ☺ ☺ ☹

Collins
New
Primary
Maths

Self assessment Unit D2

Name _____ Date _____

- I can decide what calculations to do to solve a problem and how to do them (mental methods, jottings, written methods, calculator)

- I can use a calculator to solve word problems involving decimals

- I can multiply and divide whole numbers and decimals by 10, 100 and 1000

- I can add and subtract whole numbers and decimals with two places in columns

- I can use an efficient method to multiply HTU × U, TU × TU, and U·t × U

- I can use rounding to estimate and check calculations

- I can plot co-ordinates on a grid

- I can estimate and measure angles less than 180°

- I can recognise acute, obtuse and right angles

- I can explain the difference between perimeter and area

- I can calculate the area of a shape

- I can choose and use a suitable metric unit to estimate and measure weight

- I understand weights presented as whole numbers, fractions or decimals that are the same

- I can work out the reading between two unnumbered divisions on kitchen and bathroom scales

Collins New Primary Maths

Self assessment Unit E2

Name _____ Date _____

● I can break a problem into steps and say the calculation I need to do to work out each step ☺ 😐 ☹

● I check that my answer is sensible ☺ 😐 ☹

● I can use addition and subtraction of two-digit numbers to derive sums and differences of decimals ☺ 😐 ☹

● I can recognise equivalent fractions ☺ 😐 ☹

● I can say a smaller whole number as a fraction of a larger one ☺ 😐 ☹

● I can explain how to turn a mixed number such as $2\frac{3}{4}$ into an improper fraction. I can draw a diagram to support my explanation ☺ 😐 ☹

● I can give the decimal equivalent of a simple fraction such as $\frac{3}{10}$ and explain how I know ☺ 😐 ☹

● I know that 'per cent' means 'parts in every 100', so $1\% = \frac{1}{100}$ ☺ 😐 ☹

● I can give a simple fraction such as $\frac{1}{10}$ as a percentage ☺ 😐 ☹

● I can find a simple percentage (50%, 25%, 75%, 10%) of a quantity ☺ 😐 ☹

● I can continue a sequence such as: 'There are 3 red sweets in every 10, there are 6 red sweets in every 20' ☺ 😐 ☹

● I can use division to find a unit fraction ($\frac{1}{2}$, $\frac{1}{3}$ etc.) of a number ☺ 😐 ☹

● I can double and halve two-digit numbers and explain how to use this to double and halve related decimals ☺ 😐 ☹

● I can use a calculator to solve problems and puzzles ☺ 😐 ☹

C Coll
New
Prima
Maths

Self assessment Unit A3

Name _____ Date _____

- I can choose what calculation to do when I solve problems ☺ 😐 ☹

- I can record my method for solving a problem so that I show each step. I record only what I need to, using symbols where I can ☺ 😐 ☹

- I can explain why I decided to use a particular method to solve a problem ☺ 😐 ☹

- I can say the value of each digit in a number, including decimals. I can partition a decimal in different ways ☺ 😐 ☹

- I can work out sums and differences of decimals ☺ 😐 ☹

- I can make sensible decisions about when to use a calculator ☺ 😐 ☹

- I know my tables to 10 for multiplication facts and division facts ☺ 😐 ☹

- I can find missing numbers in a sequence that contains decimals ☺ 😐 ☹

- I can use a written method to solve calculations such as HTU × U and TU × TU ☺ 😐 ☹

- I can multiply a decimal with one place by a one-digit number using a written method. I can explain each step of my calculation ☺ 😐 ☹

- I can divide a three-digit number by a one-digit number using a written method. I can explain each step of my calculation ☺ 😐 ☹

- I can estimate and check the result of a calculation ☺ 😐 ☹

C Collins New Primary Maths

Self assessment Unit B3

Name _____ Date _____

- I can split a word problem into steps and work out what calculation to do for each step. I can explain what the answer to each step tells me   ☺ 😐 ☹

- I recognise when there may be more than one solution to a problem and try to find them all   ☺ 😐 ☹

- I can suggest a general statement and test whether it is true by investigating examples   ☺ 😐 ☹

- I can add and subtract decimals in my head by using a related two-digit addition or subtraction   ☺ 😐 ☹

- Before I solve a calculation or a word problem, I work out an estimate for the answer   ☺ 😐 ☹

- I can use written methods to add and subtract whole numbers and decimals   ☺ 😐 ☹

- I can explain each step when I write addition and subtraction calculations in columns   ☺ 😐 ☹

- I can use a calculator to find missing numbers in calculations. I use inverse operations and number facts to help me   ☺ 😐 ☹

- I can recognise when to round up or down after division, depending on the context   ☺ 😐 ☹

- I can express a remainder as a fraction   ☺ 😐 ☹

- I can use tables facts to multiply multiples of 10 and 100 and find linked division facts   ☺ 😐 ☹

- I can use doubling and halving to solve related multiplication calculations   ☺ 😐 ☹

- I use mathematical vocabulary to describe the features of a 2-D shape   ☺ 😐 ☹

- I use the properties of 3-D shapes to draw their nets accurately   ☺ 😐 ☹

C Colli
New
Prima
Maths

Self assessment Unit C3

Name _____ Date _____

- I can estimate and measure capacity in litres and millilitres using appropriate measuring instruments   ☺ 😐 ☹

- I can understand capacity presented as whole numbers, fractions and decimals that are the same   ☺ 😐 ☹

- I can find the value of each interval on a scale and use this to give approximate values of readings between divisions   ☺ 😐 ☹

- I can collect and organise data to find out about a subject or to answer a question   ☺ 😐 ☹

- I can use graphs to show findings about a subject or to help explain my answer to a question   ☺ 😐 ☹

- I can decide what information needs to be collected to answer a question and how best to collect it   ☺ 😐 ☹

- I can explain what a table, graph or chart tells us and consider questions that it raises   ☺ 😐 ☹

- I can explain why I chose to represent the data using a particular table, graph or chart   ☺ 😐 ☹

- I can present and interpret data in a bar line graph   ☺ 😐 ☹

- I can present and interpret data in a line graph   ☺ 😐 ☹

- I know that the 'mode' is the most common piece of information   ☺ 😐 ☹

- I can find the mode of a set of data that I have collected   ☺ 😐 ☹

- I can describe how likely an event is to happen and justify my statement   ☺ 😐 ☹

Collins New Primary Maths

Self assessment Unit D3

Name _____ Date _____

● I can use the most efficient method of solving a problem, including using a calculator   ☺ 😐 ☹

● I can add and subtract whole numbers and decimals with up to two places in columns   ☺ 😐 ☹

● I can use efficient methods to multiply HTU × U, HT × HT, U·t × U and divide HTU by U   ☺ 😐 ☹

● I can use a calculator to solve a problem and interpret the display correctly   ☺ 😐 ☹

● I can use rounding of whole numbers and decimals to estimate and check calculations   ☺ 😐 ☹

● I can choose and use the correct metric unit to estimate and measure capacity   ☺ 😐 ☹

● I can interpret a reading between two unnumbered divisions on a scale on measuring cylinders and jugs, and read accurately the number of millilitres in a litre jug   ☺ 😐 ☹

● I can find the area of a rectangle using the formula length × breadth. I know that area is measured in units^2 ☺ 😐 ☹

● I can draw angles less than 180° to within 5° and calculate angles on a straight line   ☺ 😐 ☹

● I can complete a pattern with one or two lines of symmetry   ☺ 😐 ☹

● I can draw where a shape will be after it has been reflected or translated   ☺ 😐 ☹

● I can solve problems using a timetable written in 24-hour clock notation   ☺ 😐 ☹

C Collin
New
Primar
Maths

Self assessment Unit E3

Name _____ Date _____

● I can break a problem into steps and say the calculation I need
to do to work out each step

● I check that my answer is sensible

● I can decide and justify what calculations to do to solve a
problem and whether I will do these mentally, using a written
method or with a calculator

● I can tell you what calculations I will do to find a fraction of
a quantity

● I can give the decimal equivalent of a simple fraction such as $\frac{3}{10}$
and explain how I know

● I can find fractions that are equivalent to each other

● I know that 'per cent' means 'parts in every 100', so $1\% = \frac{1}{100}$

● I can give a simple fraction such as $\frac{1}{10}$ as a percentage

● I can tell you what calculations I will do to find a percentage of
a quantity

● I can use the relationships between numbers to solve ratio and
proportion questions

● I can use a written method to multiply HTU × U, TU × TU and
U·t × U, and explain each step

● I can use a written method to divide a three-digit number by a
one-digit number and explain each step

● I can explore patterns, identify rules and make general statements

● I can explain my thinking in words and using diagrams

Test 1
Mental mathematics test questions and answers

Say: **For this group of questions, you will have 5 seconds to work out each answer and write it down.**

	The questions	Answers
1.	What is nine times four?	36
2.	What is the sum of twenty, thirty and fifteen?	65
3.	Subtract two hundred and seventy from five hundred.	230
4.	What is half of twenty-six?	13
5.	What is the sum of one point seven and one point eight?	3·5

Say: **For the next group of questions, you will have 10 seconds to work out each answer and write it down.**

		Answers
6.	Look at your answer sheet. What is the cost of two apples and an orange?	£1.50
7.	Lisa's birthday is on the twenty-sixth of September. It is her brother's birthday nine days later. What is the date of Lisa's brother's birthday?	5th October
8.	What is two-thirds of seventy-five?	50
9.	Look at your answer sheet. What is the distance between Points A and B?	19 mm
10.	The temperature is four degrees Celsius. It falls by six degrees. What is the new temperature?	–2 °C
11.	A packet of mints cost one pound twenty-five pence each. How much do five packets of mints cost?	£6.25
12.	Look at your answer sheet. Draw a ring around the fraction that is equal to zero point four.	$\frac{2}{5}$
13.	What is four thousand subtract ten?	3990
14.	The perimeter of a square is eighty centimetres. How long is each side?	20 cm
15.	Look at your answer sheet. Draw a ring around the shape that has two lines of symmetry.	rectangle

Say: **For the next group of questions, you will have 15 seconds to work out each answer and write it down.**

		Answers
16.	Three pineapples cost three pounds seventy-five pence. What is the cost of one pineapple?	£1.25
17.	What is the difference between four hundred and ninety-five and six hundred and one?	106
18.	I multiply a number by ten. My answer is seventy-three. What was the number I started with?	7·3
19.	Subtract one point seven from two point five.	0·8
20.	What number is double two thousand, three hundred and fifty?	4700

Say: **Now put down your pencil. The test is finished.**

Test 1
Papers A and B answers

Paper A

1. 4·83

2. 4

3. a) £1.75   b) 25p

4. 349   6920
    58   78 600

5. 3·8

6.

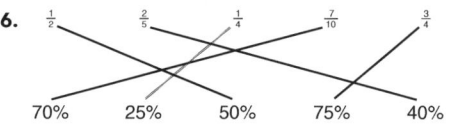

7. 22, 50, 64 and 85

8. 653

9. £8.03

10. a) 23   b) 115

11. 1:55 p.m.

12. 1:50 p.m. (or 13:50)

13. a) £17.50   b) £1.40

14. 20 cm

15. a) 7   b) 13

16. 3 km 400 m (or 3400 m)

17. 13·5 cm²

18. a) 24 m   b) £72

19. 1728

20.

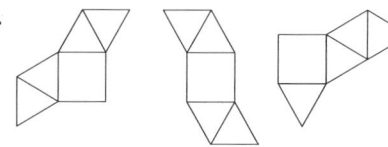

21. a) 15 °C   b) 3:00 p.m.

22. 57 mm (or 5 cm 7 mm)

23. Selma is correct. A total of 7 can occur in six ways (i.e. 1 + 6, 2 + 5, 3 + 4, 4 + 3, 5 + 2, 6 + 1). The next most common total is 6 which can occur in five ways (i.e. 1 + 5, 2 + 4, 3 + 3, 4 + 2, 5 + 1).

24. 432 cm²

25. £770

Paper B

1. 87

2. £13.80

3. 32 cm²

4. 100

5.

6. 81

7. 2·6 m

8. $\frac{1}{3} = \frac{2}{6} = \frac{4}{12} = \frac{6}{18} = \frac{10}{30}$

9. a) b)

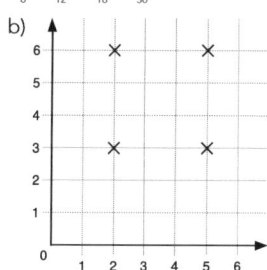

    b) (5, 6)

10. 9

11. £2.10, £5, £1.20 or £1.20, £5, £2.10

12. 20 minutes

13. 7

14. a) 13   b) tennis racket

15. 70 cm

16. 43°

17. 1·375 kg or 1 kg 375 g

18. a) 3 and 4   b) 300 and 400

19. 10th August

20. 8

21.

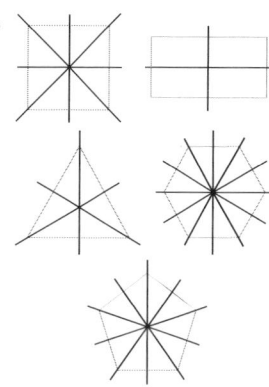

22. 45%

23. 1, 2, 3, 6, 9, 18

24.

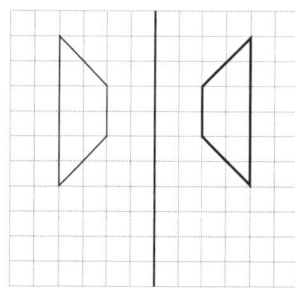

Mental mathematics test

Name ....

Date ....   Class ....

Total marks ☐

Time: 5 seconds

1		
		₁

2		20 30 15
		₂

3		270 500
		₃

4		26
		₄

5		1·7 1·8
		₅

Time: 10 seconds

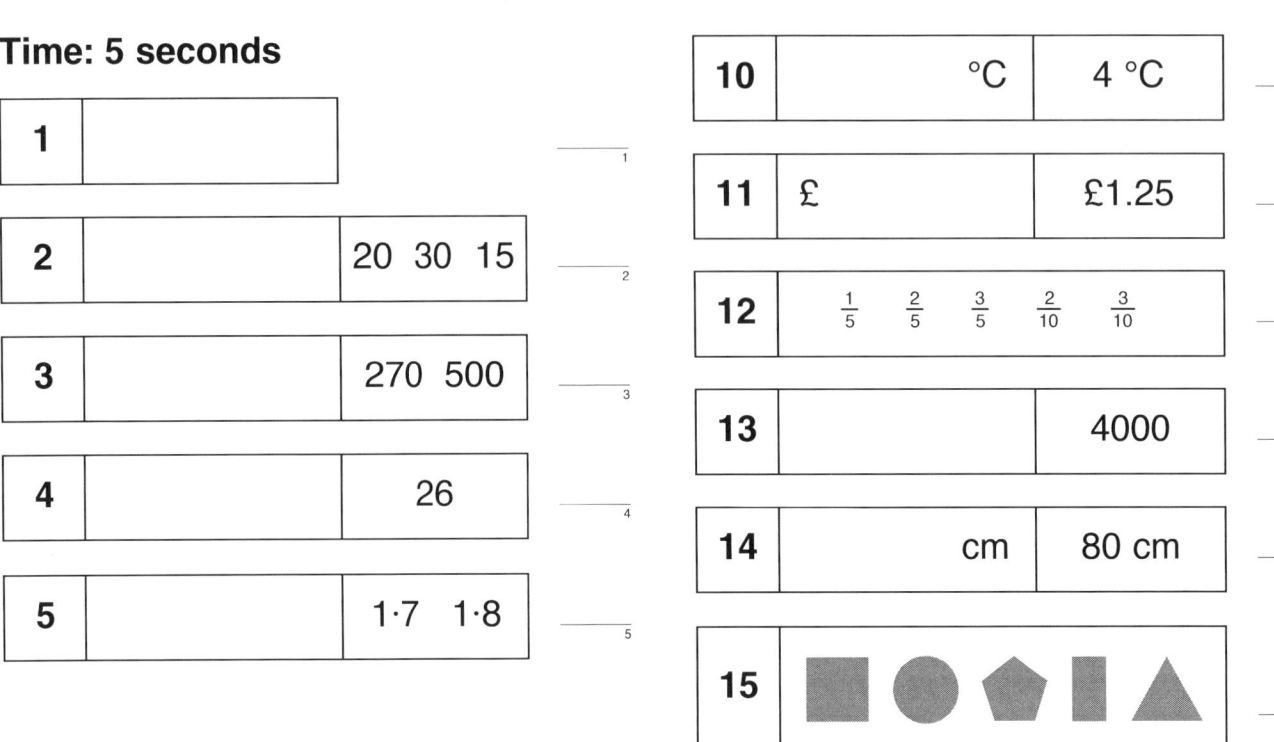

6	£	
		₆

7		26th Sept
		₇

8		75
		₈

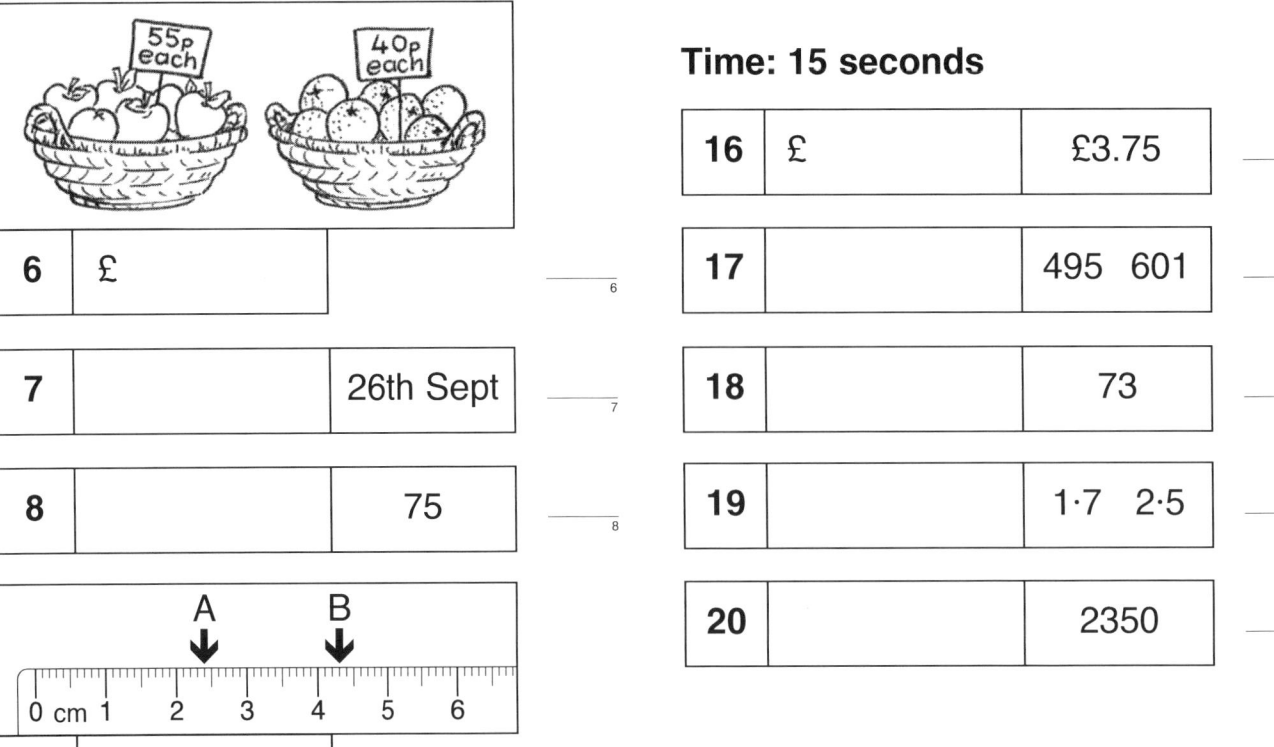

9		mm
		₉

10	°C	4 °C
		₁₀

11	£	£1.25
		₁₁

12	$\frac{1}{5}$ $\frac{2}{5}$ $\frac{3}{5}$ $\frac{2}{10}$ $\frac{3}{10}$	
		₁₂

13		4000
		₁₃

14	cm	80 cm
		₁₄

15	■ ● ⬠ ▮ ▲	
		₁₅

Time: 15 seconds

16	£	£3.75
		₁₆

17		495 601
		₁₇

18		73
		₁₈

19		1·7 2·5
		₁₉

20		2350
		₂₀

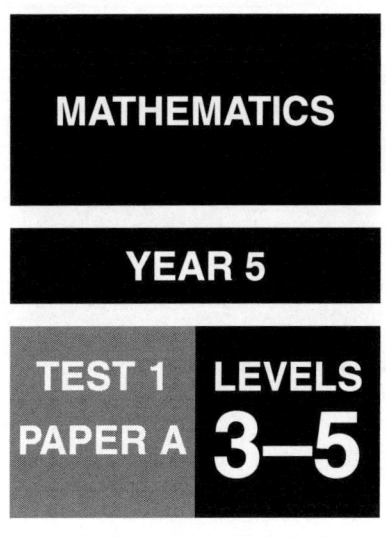

MATHEMATICS

YEAR 5

TEST 1 PAPER A **LEVELS 3–5**

CALCULATOR NOT ALLOWED

PAGE	MARKS
3	
4	
5	
6	
7	
8	
9	
10	
11	
12	
TOTAL	

RESOURCES
- pencil

Name

Date

Class

Instructions

You may not use a calculator to answer any questions in this test.

Work as quickly and as carefully as you can.

You have 45 minutes for this test.

If you cannot do one of the questions, go on to the next one.

You can come back to it later if you have time.

If you have finished before the end, go back and check your work.

Follow the instructions for each question carefully.

 This shows where you need to put your answer.

If you need to do any working out, you can use any space on the page.

Some questions have an answer box like this:

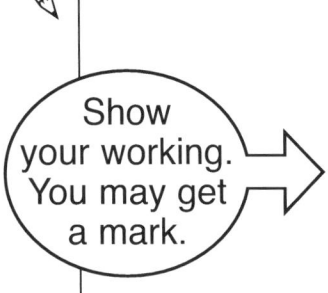

Show your working. You may get a mark.

For these questions you may get a mark for showing your working.

1 Circle the decimal closest to 5.

4·47   5·38   5·51   4·83   3·55

—————
1 mark

2 Write in the missing digit.

$$\begin{array}{r} 6\ \square \\ \times\quad 7 \\ \hline 4\ 4\ 8 \\ \hline \end{array}$$

—————
1 mark

3 An apple costs 35p.

a) What is the cost of 5 apples?

3a

—————
1 mark

b) How much change would you get from £2?

3b

—————
1 mark

4 Multiply each of the following numbers by 100.

 3·49    69·2

0·58    786

4

—————
1 mark

5 Write the missing number on this number line.

[   ]   4   4·2   4·4

5

1 mark

6 Draw a line between each fraction and its equivalent percentage.
One has been done for you.

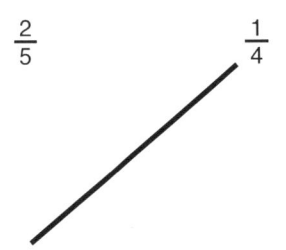

$\frac{1}{2}$ $\frac{2}{5}$ $\frac{1}{4}$ $\frac{7}{10}$ $\frac{3}{4}$

70%   25%   50%   75%   40%

6

1 mark

7 Circle all the numbers that are 1 more than a multiple of 7.

22   39   50   64   73   85

7

1 mark

8 Circle the number that has 5 tens, 6 hundreds and 3 units.

365   563   65·3   5·63   653

8

1 mark

9 Calculate £5.60 + £2.43

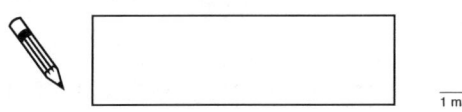

9

1 mark

10 The table shows the number of pens and pencils for sale in a shop.

	black	coloured
pens	23	31
pencils	47	68

a) How many pens are black?

10a

1 mark

b) How many pencils are for sale?

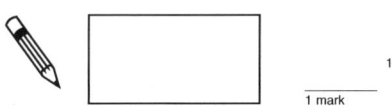

10b

1 mark

11 Circle the time that is the same as thirteen fifty-five.

3:55 p.m.   1:55 a.m.   5:13 p.m.

1:55 p.m.   3:55 a.m.

11

1 mark

12 Tim caught a train at 10:20 a.m.
The journey took 3 hours 30 minutes.
At what time did the train arrive?

12

1 mark

13 Five children went to the cinema. Their tickets cost £3.50 each.

a) How much did their tickets cost altogether?

13a

1 mark

b) They had £7 between them to spend equally on food and drink. How much did each of them get to spend on food and drink?

13b

1 mark

14 This shape is made up of 1 centimetre squares.
What is its perimeter?

This diagram
is not to scale.

14

1 mark

15 This graph shows the number of hours that the children in a class spent on a computer in one week.

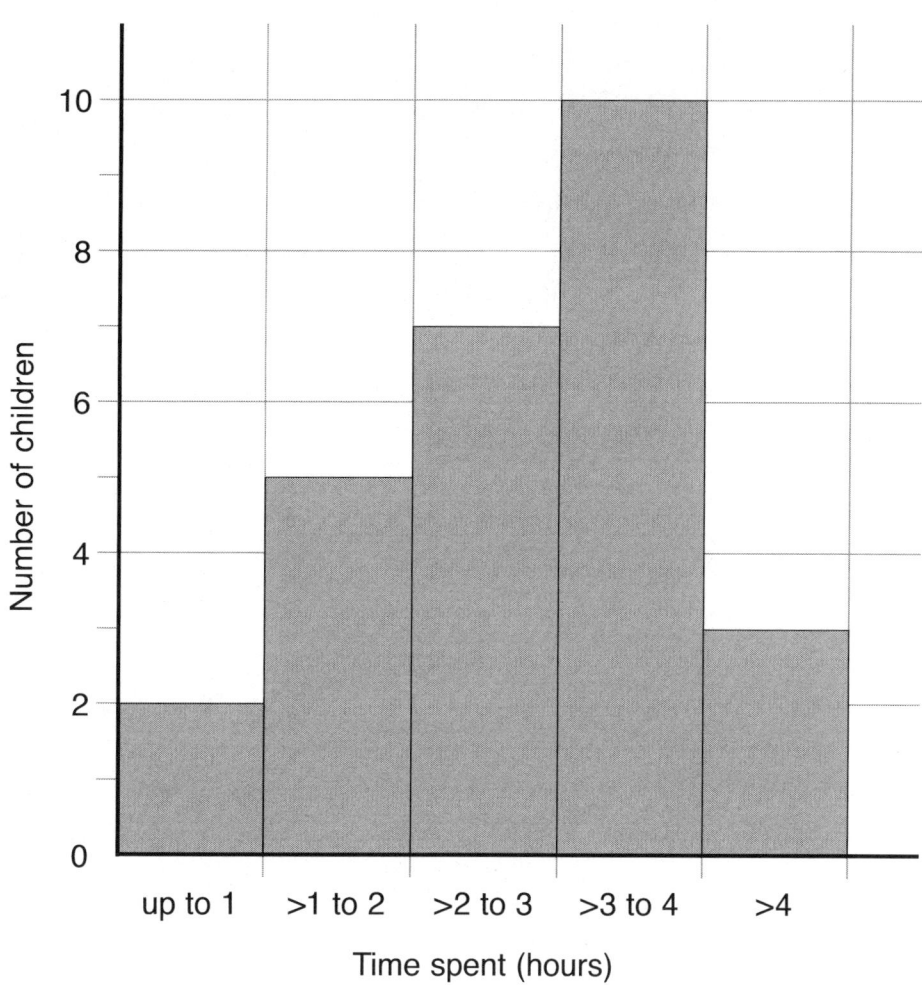

a) How many children spent 2 hours or less on the computer?

15a

1 mark

b) How many children spent more than 3 hours on the computer?

15b

1 mark

16 Sonni lives 1 km 700 m from his school. He walks there and back each day. How far is this?

16

1 mark

17 What is the area of this shape?

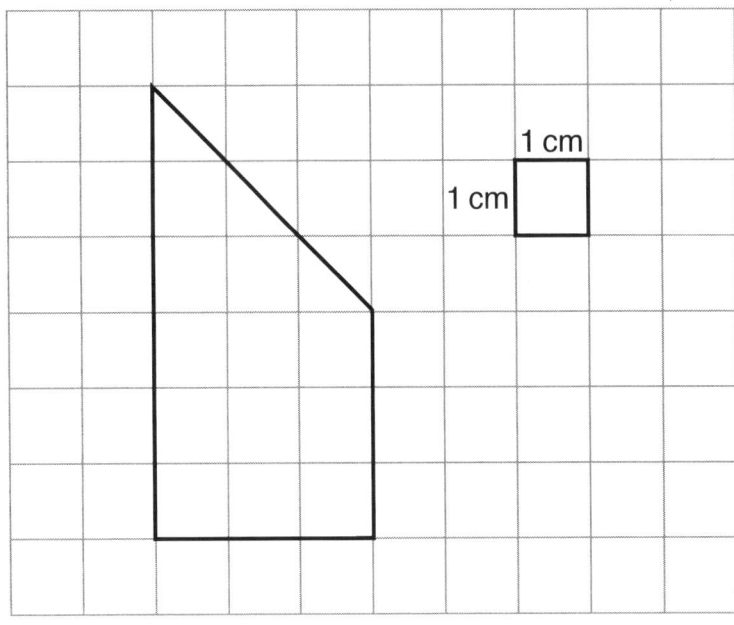

This diagram is not to scale.

1 cm

1 cm

17

1 mark

18 Sanjay is making 8 curtains. He uses 3 m of material to make each one.

a) How much material does he need?

18a

1 mark

b) The material costs £3 a metre.
   How much does he spend on material?

18b

1 mark

19 Boxes of baked beans are packed with 4 layers of tins, each layer with 36 tins in it. Mr Greg orders 12 boxes of baked beans for his shop. How many tins of beans is this?

Show your working. You may get a mark.

19

2 marks

20 Draw a cross by the figure that is **not** the net of a squared-based pyramid.

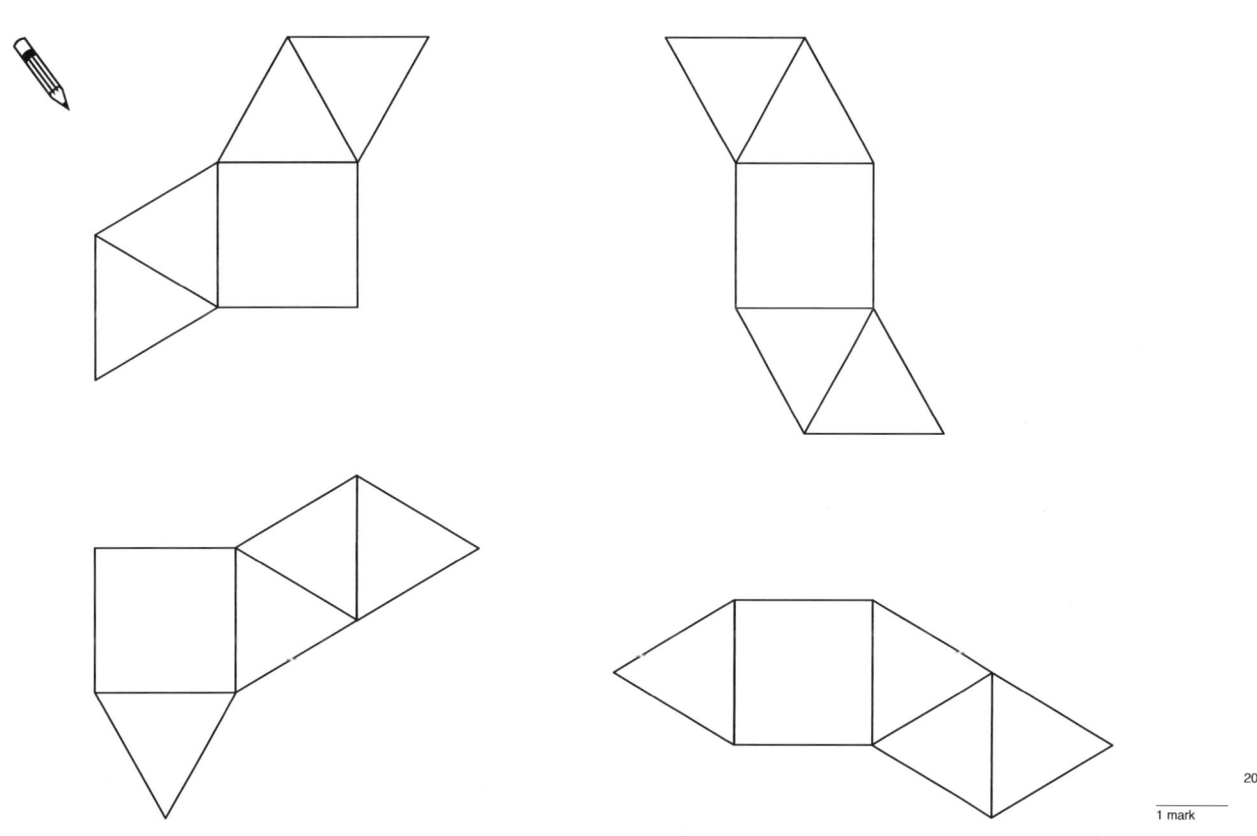

20

1 mark

21 This graph shows the temperature on a day in August.

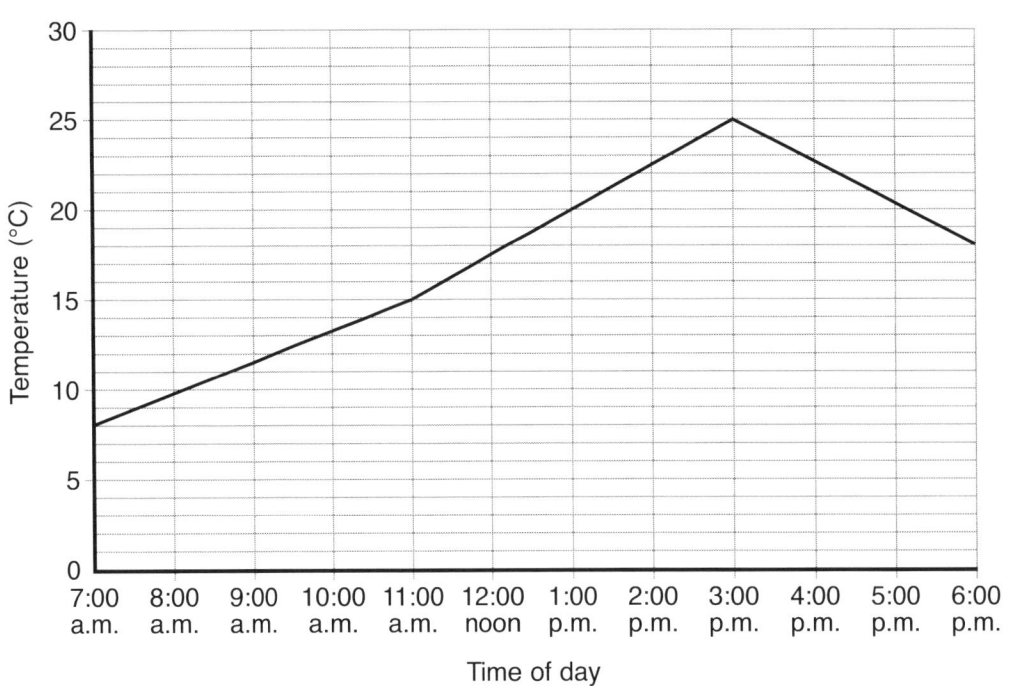

a) What was the temperature at 11:00?

21a

1 mark

b) What was the hottest time of day?

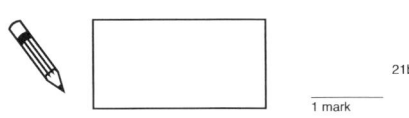

21b

1 mark

22 What is the distance between point A and point B?

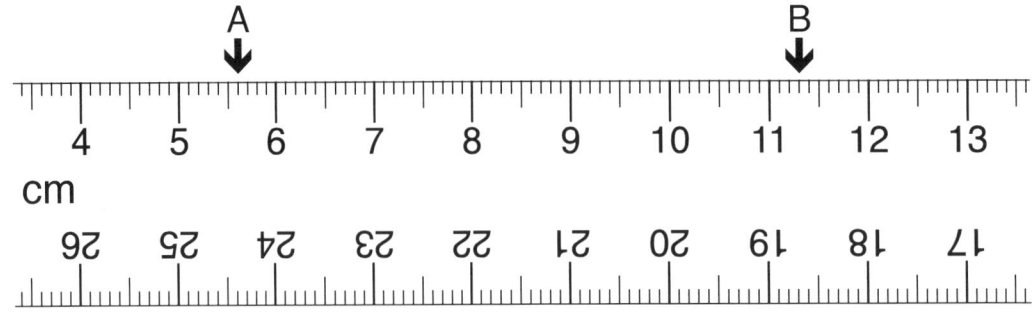

22

1 mark

23 Tom rolls two 1–6 dice.

Selma says that when Tom adds the two scores together he is more likely to get a total of 7 than any other possible total.

Is she correct?

Explain why you think this.

2 marks

24 What is the area of this rectangle?

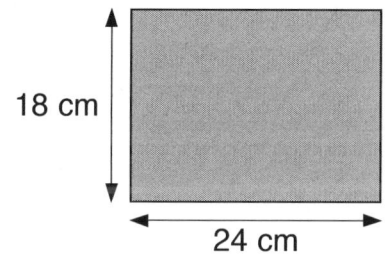

18 cm

24 cm

This diagram is not to scale.

Show your working. You may get a mark.

cm²

2 marks

25 Anna earns £2200 each month. She spends 35% of this on rent. How much does she spend on rent?

Show your working. You may get a mark.

End of test

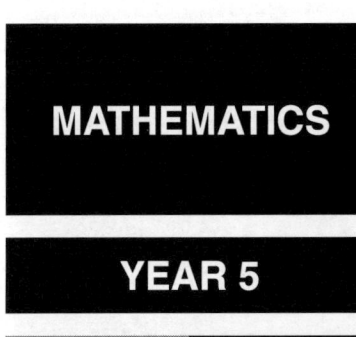

MATHEMATICS

YEAR 5

TEST 1
PAPER B

LEVELS 3–5

CALCULATOR
ALLOWED

PAGE	MARKS
3	
4	
5	
6	
7	
8	
9	
10	
11	
12	
TOTAL	

RESOURCES

- pencil
- ruler
- calculator

Name

Date

Class

Instructions

You may use a calculator to answer any questions in this test.

Work as quickly and as carefully as you can.

You have 45 minutes for this test.

If you cannot do one of the questions, go on to the next one.

You can come back to it later if you have time.

If you have finished before the end, go back and check your work.

Follow the instructions for each question carefully.

 This shows where you need to put your answer.

If you need to do any working out, you can use any space on the page.

Some questions have an answer box like this:

Show
your working.
You may get
a mark.

For these questions you may get a mark for showing your working.

1 Write in the missing number.

$362 + \boxed{} = 449$

1 mark

---

2 Toni has a note and some coins.

How much money does Toni have?

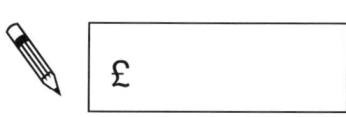 £ $\boxed{}$

2

1 mark

---

3 Here is a shape drawn on a grid of centimetre squares.

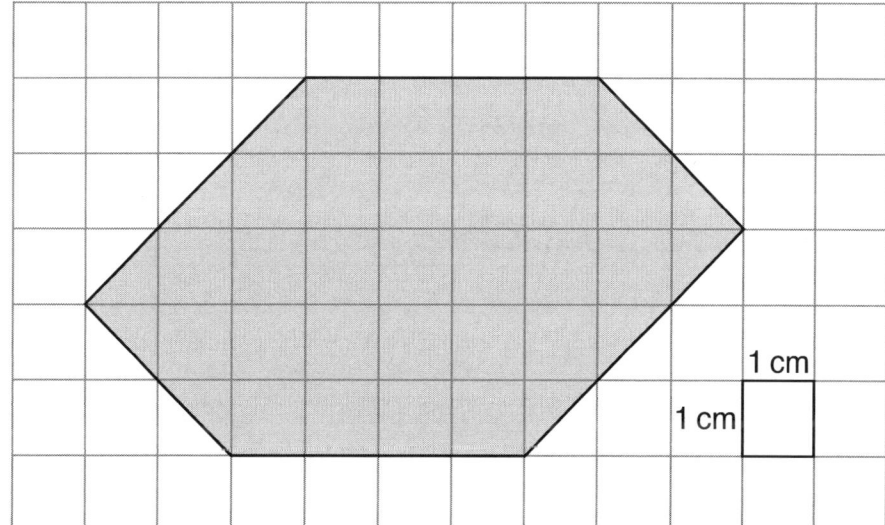

What is the area of this shape?

 $\boxed{}$ cm^2

3

1 mark

---

4   Write in the missing number.

$$\frac{364}{\boxed{}} = 3{\cdot}64$$

1 mark

5   Here is part of a number line.

Draw an arrow (↓) to show the position of 16.

5
1 mark

6   Circle the number that is **not** a common multiple of 6 and 9.

18   54   36   81   72

6
1 mark

7   What is the distance between A and B?

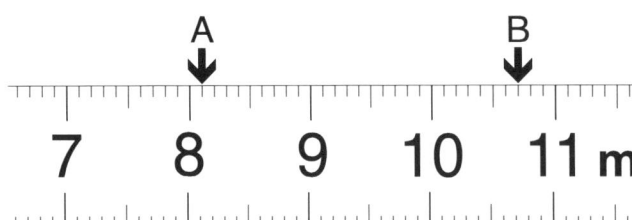

7
1 mark

8 Write a number in each box to make these fractions equivalent.

$$\frac{1}{3} = \frac{2}{\boxed{}} = \frac{\boxed{}}{12} = \frac{\boxed{}}{18} = \frac{10}{\boxed{}}$$

8

2 marks

9 The co-ordinates of three points of a square are:

(2, 3), (2, 6) and (5, 3).

a) Mark these points on the grid.

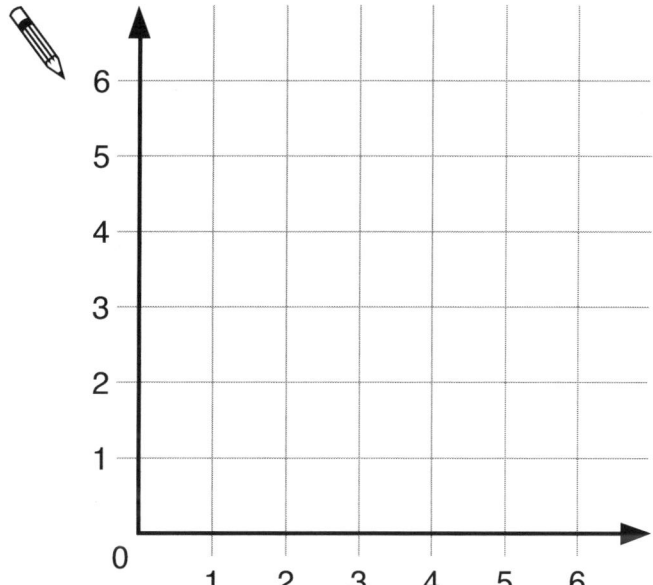

9a

1 mark

b) Now mark the fourth corner of the square.

9b

1 mark

The co-ordinates of the fourth corner are: (____ , ____)

10 Calculate 15% of 60.

10

1 mark

11 Here are some amounts of money.

£1.20   £5   £2.10

Write one of these amounts in each box to make this number story correct.

Fiona buys a notebook for £1.70 and a packet

of crayons for [   ]. She pays with a

[   ] note and receives [   ] change.

12 Ana had a 50 minute tennis lesson starting at 9:30 a.m. Afterwards she walked straight home. She arrived home at 10:40 a.m.

How long did it take her to walk home?

Show your working. You may get a mark.

13 Write in the missing number.

$634 \times$ [   ] $= 4438$

14 A group of children were asked what sports equipment they owned. The table shows the results.

Type of equipment	Number of children
Skateboard	19
Rollerblades	15
Football boots	28
Tennis racket	8
Swimming trunks	43

a) How many more children own football boots than rollerblades?

14a

1 mark

b) What piece of sports equipment do 11 fewer children have than a skateboard?

14b

1 mark

15 Rashid has some bricks like this.

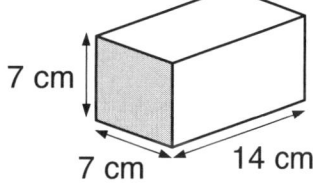

7 cm

7 cm   14 cm

He builds this wall with 7 of them.
How long is the wall?

Show your working. You may get a mark.

15

2 marks

16 Here is a triangle.

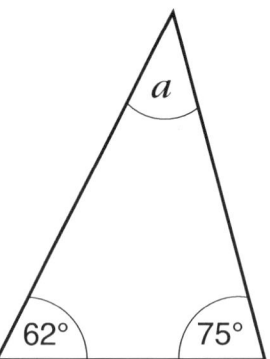

Not to scale

Calculate the size of angle a.

Do not use a protractor (angle measurer).

Show your working. You may get a mark.

16

2 marks

17 Sal has $2\frac{1}{2}$ kg of flour. She uses $\frac{1}{4}$ of it to make some bread and $\frac{1}{5}$ of it to make a cake. How much flour does she have left?

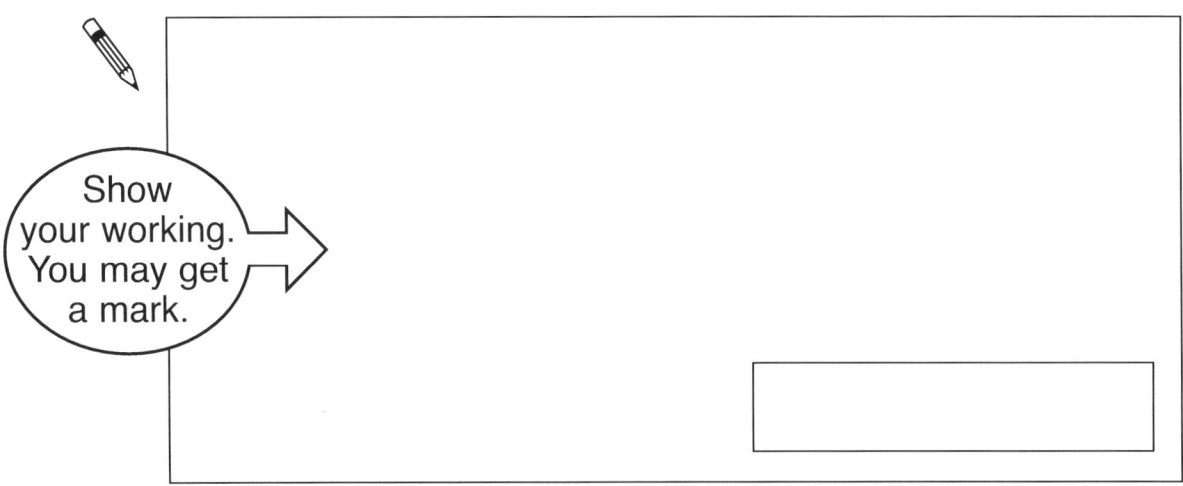

Show your working. You may get a mark.

17

2 marks

18

a is a number between 30 and 40.

Complete the following sentences.

One has been done for you.

a – 10 is a number between 20 and 30.

 a) $\frac{a}{10}$ is a number between ☐ and ☐ .

18a

1 mark

 b) *a* × 10 is a number between ☐ and ☐ .

18b

1 mark

19

Here is the calendar for July when Ismael went to stay with his aunt.

July					
Sunday	June 29	6	13	20	27
Monday	June 30	7	14	21	28
Tuesday	1	8	15	22	29
Wednesday	2	9	16	23	30
Thursday	3	10	17	24	31
Friday	4	11	18	25	August 1
Saturday	5	12	19	26	August 2

Ismael went to stay with his aunt on 27th July.

He stayed 14 nights with her and then came home.

On what date did he come home?

☐

19

1 mark

20 $2\frac{1}{3} + 5\frac{2}{3} =$

20

1 mark

21 Draw the lines of symmetry on each of these shapes.

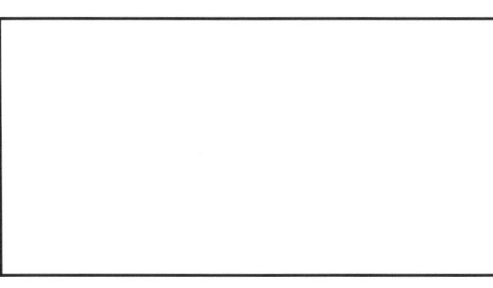

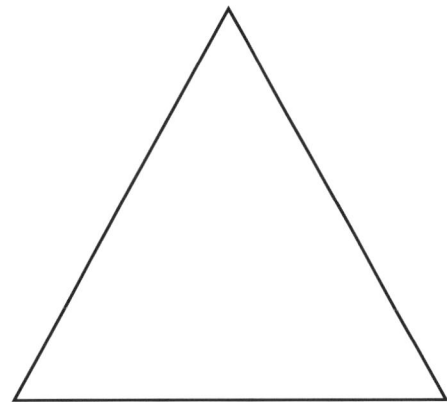

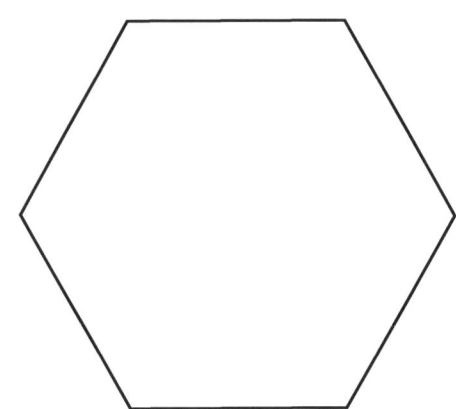

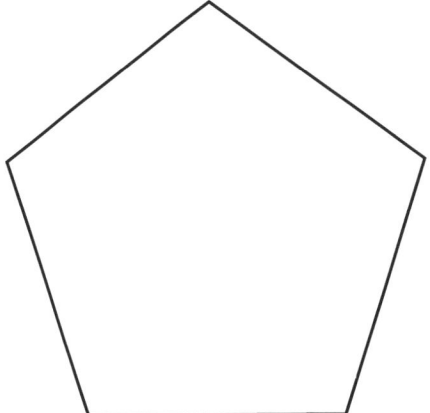

21

2 marks

22 At Sally's birthday party her birthday cake was cut into 20 pieces.

This is the cake after the party.

What percentage of the cake has been eaten?

Show your working. You may get a mark.

22

2 marks

23 Write all the factors of 18.

23

2 marks

24 Reflect the shape along the line of symmetry.

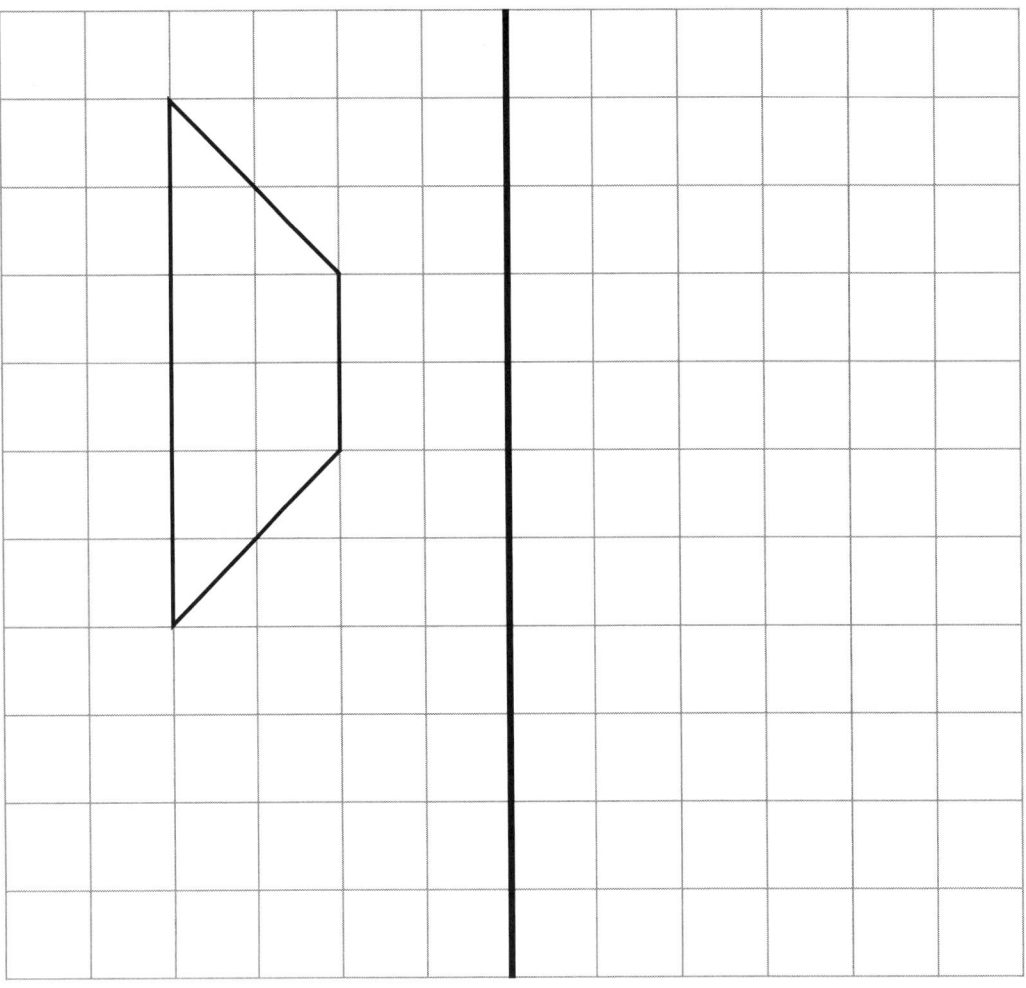

End of test

Test 2
Mental mathematics test questions and answers

Say: **For this group of questions, you will have 5 seconds to work out each answer and write it down.**

	The questions	Answers
1.	What is double zero point four two?	0·84
2.	What is fifty-six divided by seven?	8
3.	What is the sum of two point three and five point six?	7·9
4.	What is the difference between four hundred and two hundred and forty?	160
5.	What is forty multiplied by thirty?	1200

Say: **For the next group of questions, you will have 10 seconds to work out each answer and write it down.**

		Answers
6.	Tiger bars cost forty-five pence each. How much do three Tiger bars cost?	£1.35
7.	What is two-thirds of ninety?	60
8.	William's trip starts at one thirty p.m. It takes eight and a half hours. At what time does his trip end?	10:00 p.m.
9.	Look at your answer sheet. Draw a ring around the decimal that is equal to three-fifths.	0·6
10.	How many vertices does a tetrahedron have?	4
11.	The area of a square is thirty-six square centimetres. How long is each side?	6 cm
12.	What is one thousand minus one hundred and ten?	890
13.	The temperature is minus five degrees Celsius. It rises by seven degrees. What is the new temperature?	2 °C
14.	Look at your answer sheet. The scale shows the weight of some rice. What is the weight of rice?	750 g
15.	The perimeter of a regular octagon is forty centimetres. What is the length of each side?	5 cm

Say: **For the next group of questions, you will have 15 seconds to work out each answer and write it down.**

		Answers
16.	Add four point six to five point eight.	10·4
17.	What number is half of six thousand, two hundred and ninety?	3145
18.	What number is one hundred and ninety-nine more than five hundred and thirty-eight?	737
19.	I multiply a number by one thousand. My answer is two thousand four hundred. What was the number I started with?	2·4
20.	A chocolate bar costs sixty pence. How many chocolate bars can I buy with ten pounds?	16

Say: **Now put down your pencil. The test is finished.**

Test 2
Papers A and B answers

Paper A

1. 466·5
2.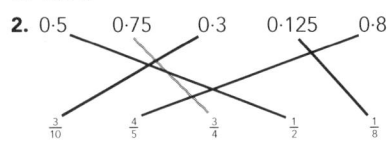
   0.5   0.75   0.3   0.125   0.8

   $\frac{3}{10}$ $\frac{4}{5}$ $\frac{3}{4}$ $\frac{1}{2}$ $\frac{1}{8}$

3. 34, 52, 76
4. 70   860
5. 3
6. 6·87   0·758
   79·4   2·439
7. a) 72   b) swallow
8. 16:25
9. a) £1.95   b) £3.05
10. 7·68
11. £4.13
12. a) 27   b) 179
13. 5:10 p.m.
14. 4
15. 20 cm
16. 47p
17. 16 cm²
18. a) 60 kg   b) £9.60
19. 9504
20.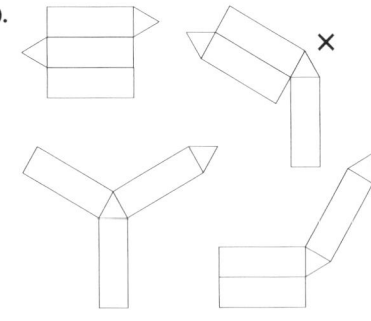
21. a) 10 km per hour   b) 25 minutes
22. 1 kg 350 g (or 1.35 kg)
23.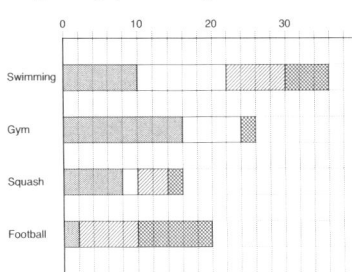

24.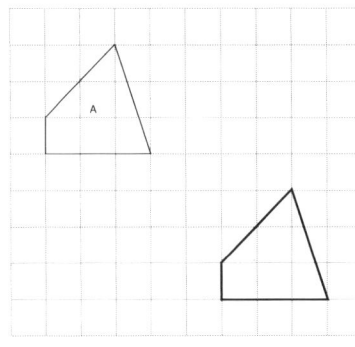

25. 3·2 kg (or 3200 g)

Paper B

1.
   0   50
2.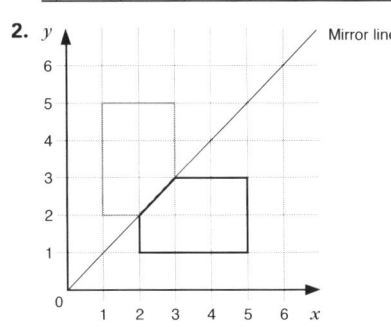
   (2,1), (5,1) and (5,3)
3. 24·5
4. £12.90
5. $\frac{1}{4} = \frac{3}{12} = \frac{4}{16} = \frac{6}{24} = \frac{8}{32}$
6. 35, 8, 4, 3 or 35, 4, 8, 3
7. 1, 2, 3, 4, 6, 8, 12, 24
8. a) 20   b) 25
9. 4
10. 7
11. 350 g
12. 16 cm²
13. 31
14. 818
15. a) 26   b) 10
16. 53°
17. a) 3·2 km (or 3 km 200 m or 3200 m)
    b) 12 minutes

18. 5:25 p.m.
19. 100
20.

	No. of lines of symmetry	No. of corners
semi-circle	1	2
square	4	4
rectangle	2	4
equilateral triangle	3	3
regular hexagon	6	6

21.

	No. of children
Cat	☺ ☺ ☺ ☺
Dog	☺ ☺ ☺
Fish	☺ ☺ ☺ ☺ ☾
Bird	☺ ☾
None	☺ ☺ ☺ ☺ ☺

22. 42
23. a)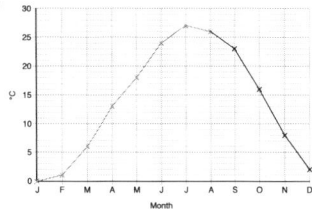

    b) A line graph illustrates the rising and falling temperature better than a bar chart because the intermediary intervals have meaning.

24.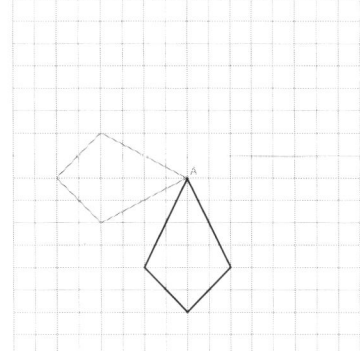

Mental mathematics test

Name ....

Date .... Class ....

Total marks ☐

Time: 5 seconds

1		0·42		1
2		56		2
3		2·3 5·6		3
4		400 240		4
5		40 30		5

Time: 10 seconds

6	£	45p		6
7		90		7
8		1:30 p.m.		8
9	0·3 0·4 0·5 0·6 0·7		9
10				10
11	cm	36 cm²		11
12		1000 110		12

| 13 | | °C | −5 °C | | 13 |

| 14 | | g | | 14 |

| 15 | cm | 40 cm | | 15 |

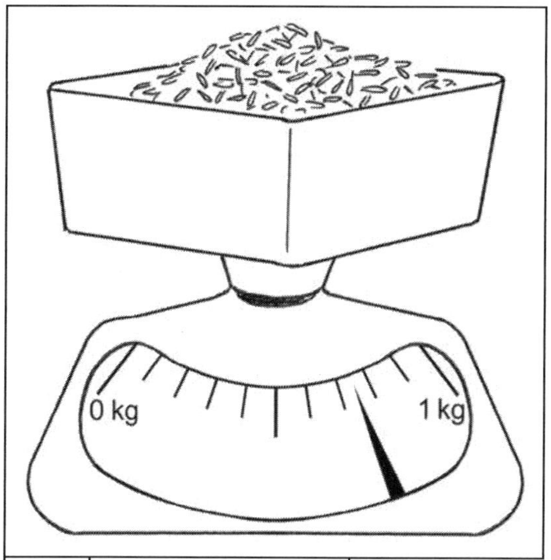

Time: 15 seconds

16		4·6 5·8		16
17		6290		17
18		199 538		18
19		2400		19
20		60p £10		20

Year 5 Test 2

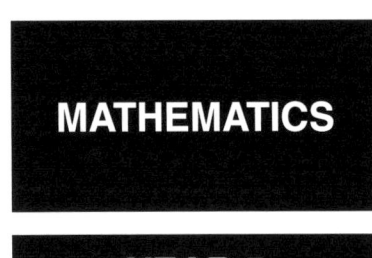

MATHEMATICS

YEAR 5

TEST 2 PAPER A | LEVELS 3–5

CALCULATOR NOT ALLOWED

PAGE	MARKS
3	
4	
5	
6	
7	
8	
9	
10	
11	
12	
TOTAL	

RESOURCES
- pencil
- ruler

Name

Date

Class

Instructions

You may not use a calculator to answer any questions in this test.

Work as quickly and as carefully as you can.

You have 45 minutes for this test.

If you cannot do one of the questions, go on to the next one.

You can come back to it later if you have time.

If you have finished before the end, go back and check your work.

Follow the instructions for each question carefully.

 This shows where you need to put your answer.

If you need to do any working out, you can use any space on the page.

Some questions have an answer box like this:

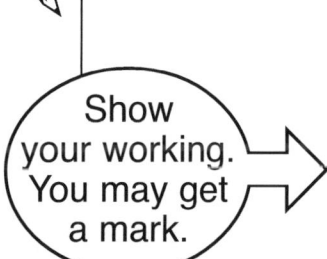

Show your working. You may get a mark.

For these questions you may get a mark for showing your working.

1 Here is a number line.

Write the number shown by the arrow.

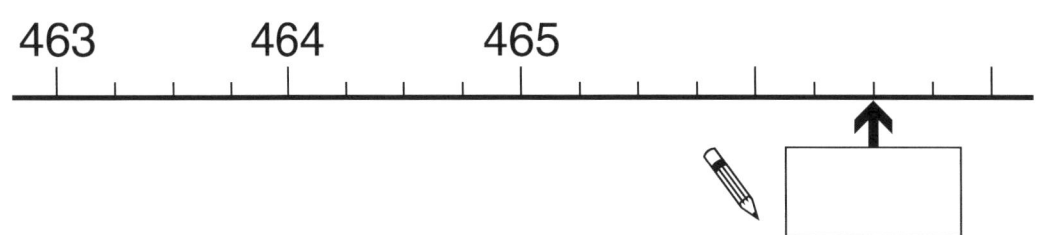

463   464   465

1 mark

2 Draw a line between each decimal and its equivalent fraction.

One has been done for you.

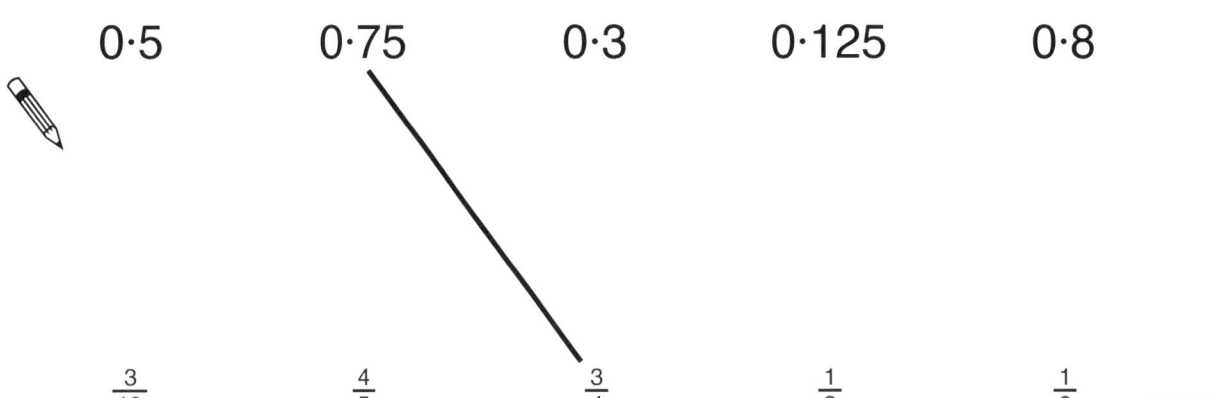

0·5   0·75   0·3   0·125   0·8

$\frac{3}{10}$ $\frac{4}{5}$ $\frac{3}{4}$ $\frac{1}{2}$ $\frac{1}{8}$

1 mark

3 Circle all the numbers that are 2 less than a multiple of 6.

 18   34   44   52   66   76

1 mark

4 Round the following numbers to the nearest multiple of ten.

 65    864

1 mark

5 Write in the missing digit.

$2 \boxed{} 4 \div 6 = 39$

1 mark

6 Divide each of the following numbers by 10.

68·7 [ ]

7·58 [ ]

794 [ ]

24·39 [ ]

1 mark

7 The children in Year 5 voted for their favourite bird.

Favourite bird	Number of children
Robin	19
Duck	3
Sparrow	7
Eagle	14
Swallow	18
Blackbird	11

a) How many children are there in Year 5?

[ ]

1 mark

b) 7 more children voted for one of the birds than voted for the blackbird. Which bird is this?

[ ]

1 mark

Total out of 4 _____

8   Show the time 4:25 in the afternoon on this 24-hour digital clock.

8

1 mark

9   Malek buys 3 pens for 65p each and pays with a £5 note.

a) How much does Malek spend altogether?

9a

1 mark

b) How much change does he get from £5?

9b

1 mark

10   Circle the decimal closest to 8.

   7·28   6·98   8·41   7·68   8·34

10

1 mark

11   Calculate £7.40 − £3.27

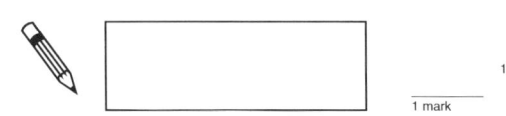

11

1 mark

12 This diagram shows the bikes for sale in a shop.

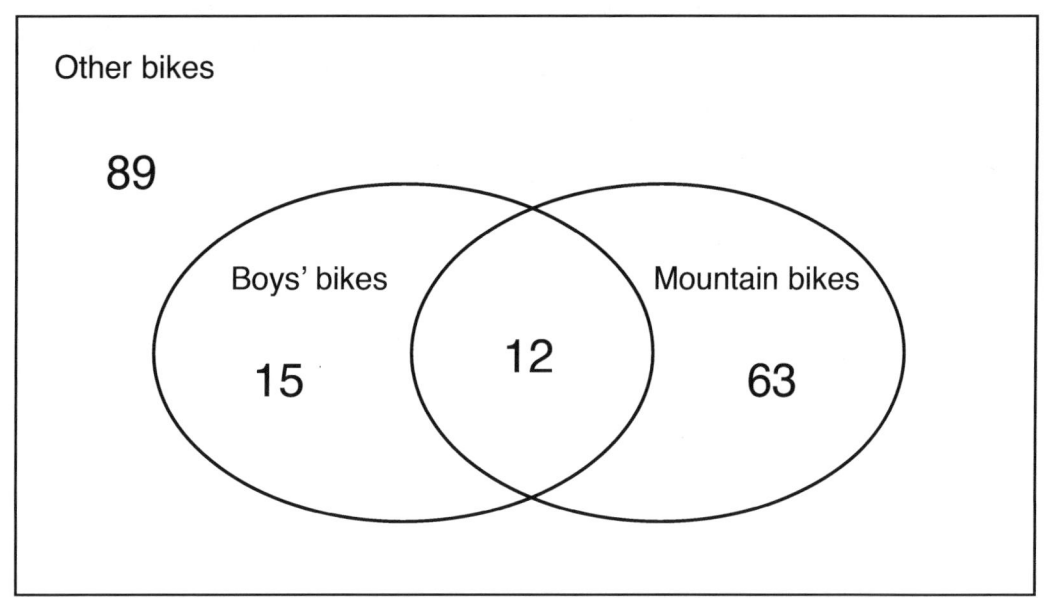

a) How many boys' bikes are for sale?

12a

1 mark

b) How many bikes are for sale altogether?

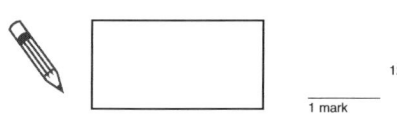

12b

1 mark

13 A film starts at 2:30 p.m. and last for 2 hours 40 minutes.
At what time does it finish?

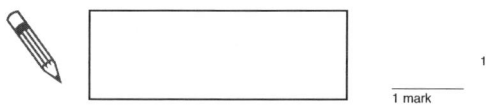

13

1 mark

14 Enid uses 750 g of flour to make a cake.
How many cakes can she make with a 3 kg bag of flour?

14

1 mark

15 This shape is made up of 1 centimetre squares.

What is its perimeter?

This diagram is not to scale.

15

1 mark

16 Salma bought five packets of crisps with a £5 note and got £2.65 in change.

How much did each packet of crisps cost?

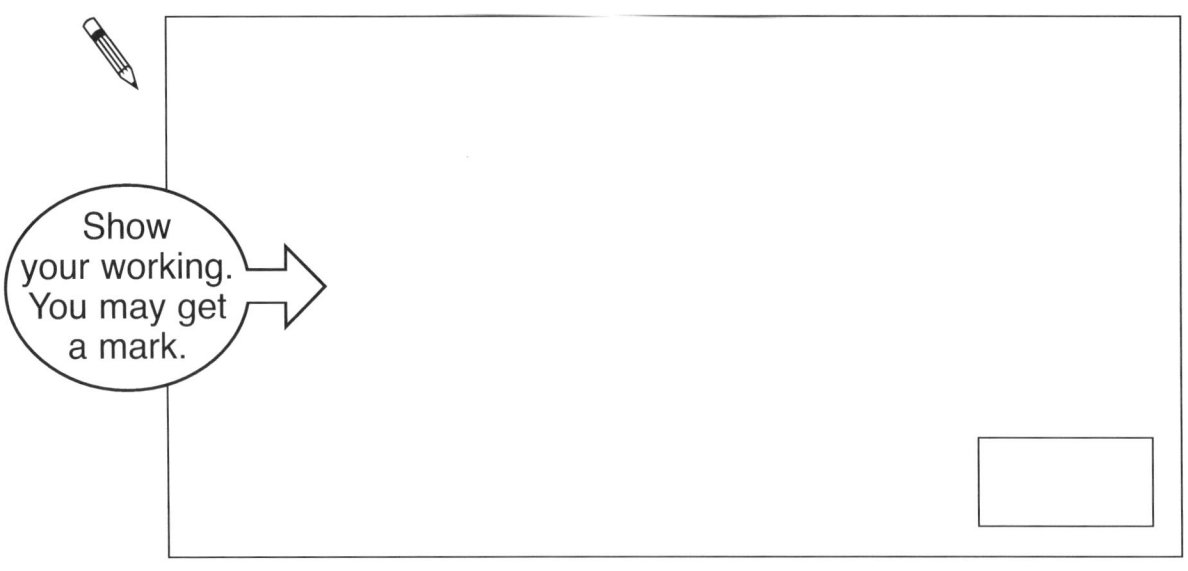

Show your working. You may get a mark.

16

2 marks

17 What is the area of this shape?

This diagram is not to scale.

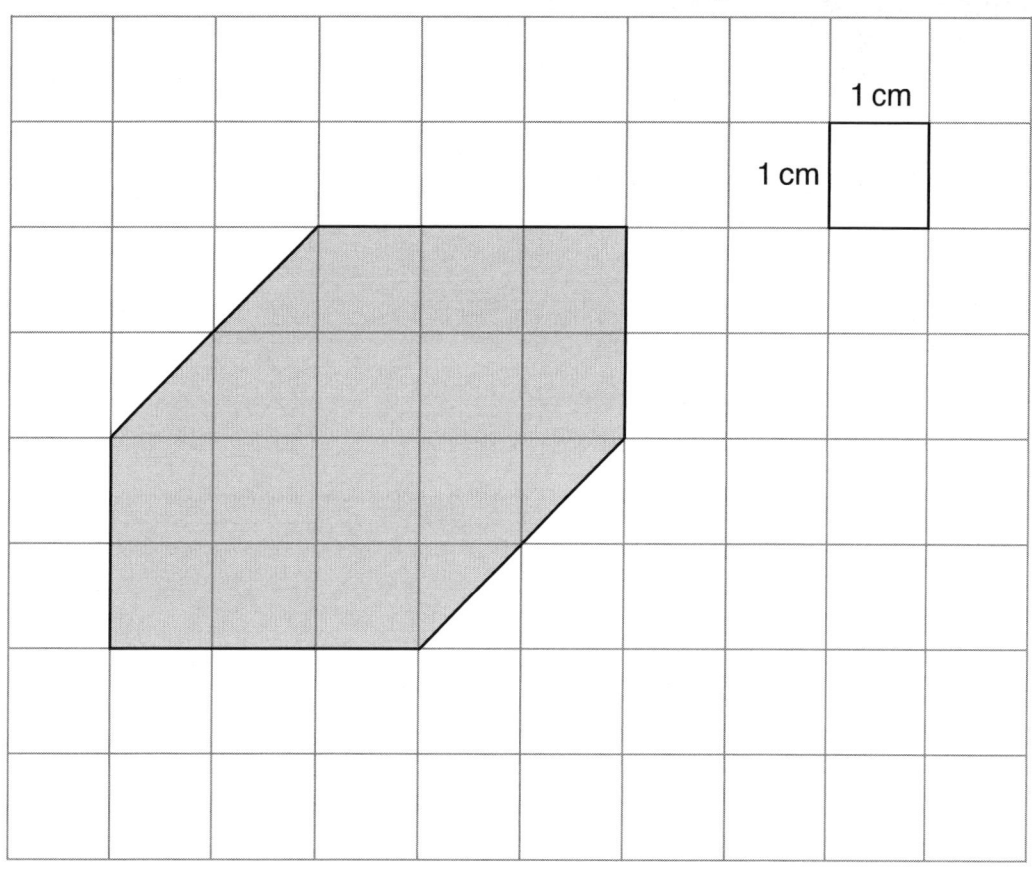

1 cm

1 cm

17

1 mark

18 Fatima is repairing a wall. She needs 24 bricks.

a) If each brick weighs $2\frac{1}{2}$ kg, what weight of bricks must she buy?

18a

1 mark

b) The bricks cost 40p each. How much does she spend on the bricks?

18b

1 mark

19 Mrs Shim buys 18 boxes of crayons for her school. Each box contains 44 packets of crayons, with 12 different colours in each packet. How many crayons does she buy altogether?

Show your working. You may get a mark.

19

2 marks

20 Draw a cross by the figure that is **not** the net of a triangular prism.

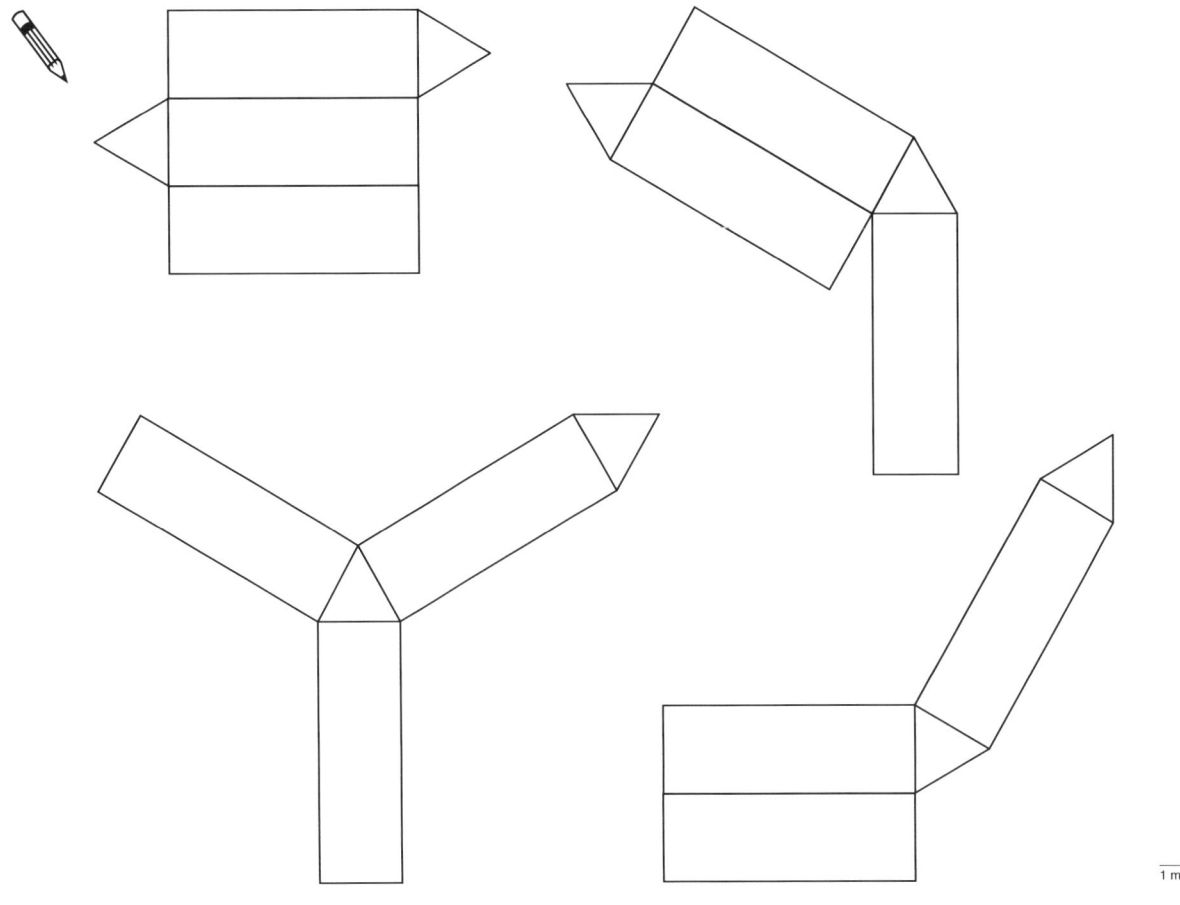

20

1 mark

21 This graph shows the speed of a ferry as it crosses to an island.

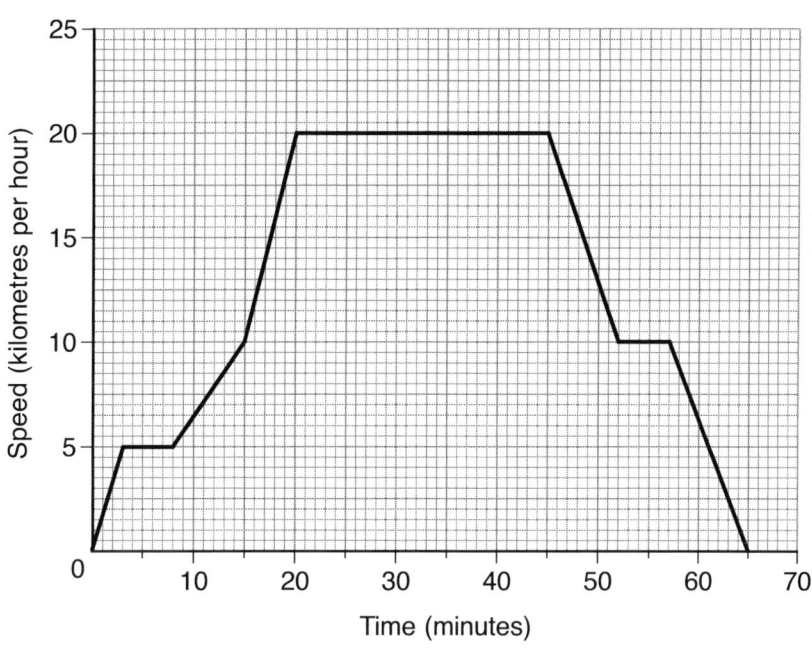

a) How fast was the ferry going after 15 minutes?

21a

1 mark

b) For how long did the ferry travel at its fastest speed?

21b

1 mark

22 What weight of flour is on the scales?

22

1 mark

23 These are the results of a survey of users of a sports centre.

	Swimming	Gym	Squash	Football
Men	10	16	8	2
Women	12	8	2	0
Girls	8	0	4	8
Boys	6	2	2	10

Show these results on the graph. Two activities have been filled in for you.

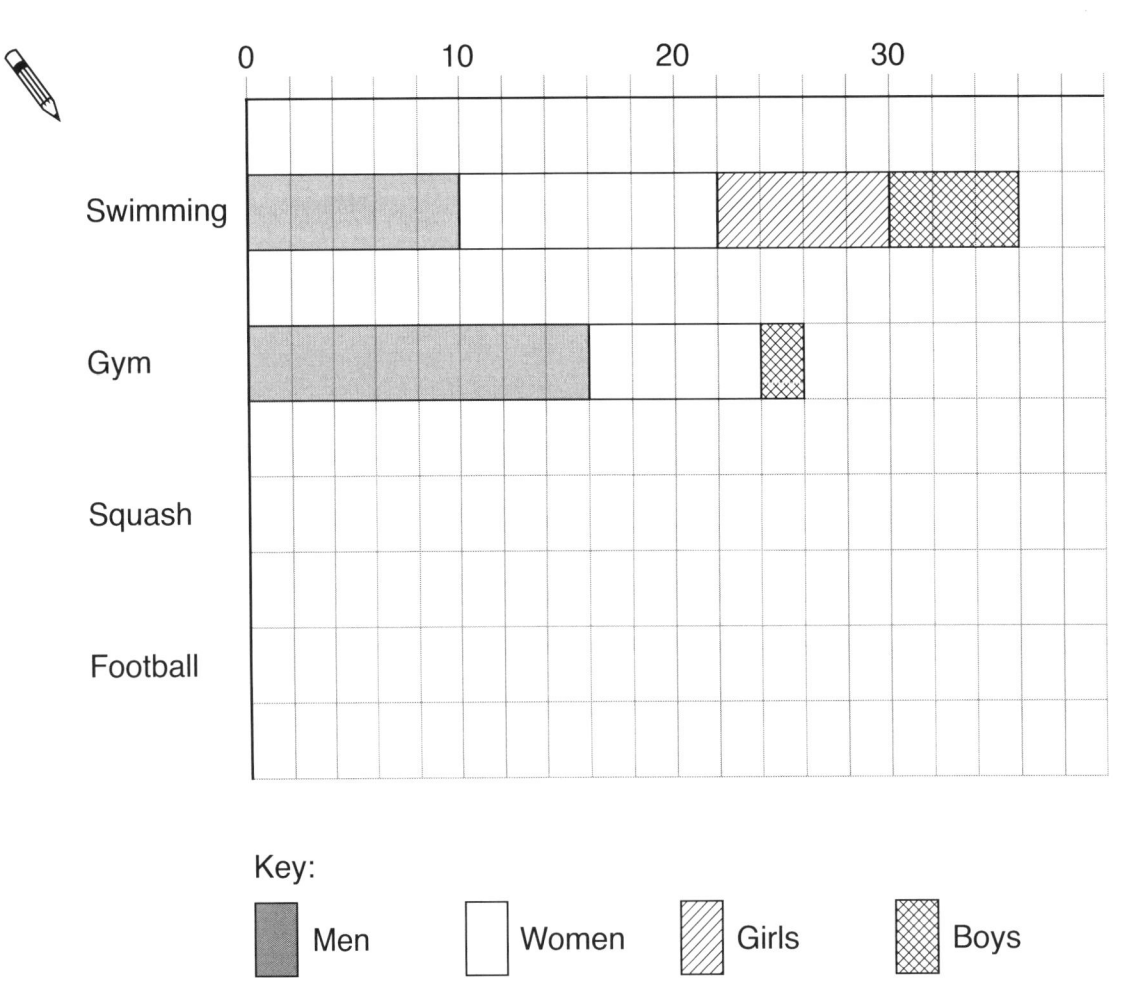

Key:

Men   Women   Girls   Boys

23

2 marks

Total out of 2 _____

24 Translate shape A 4 squares down and 5 squares to the right.

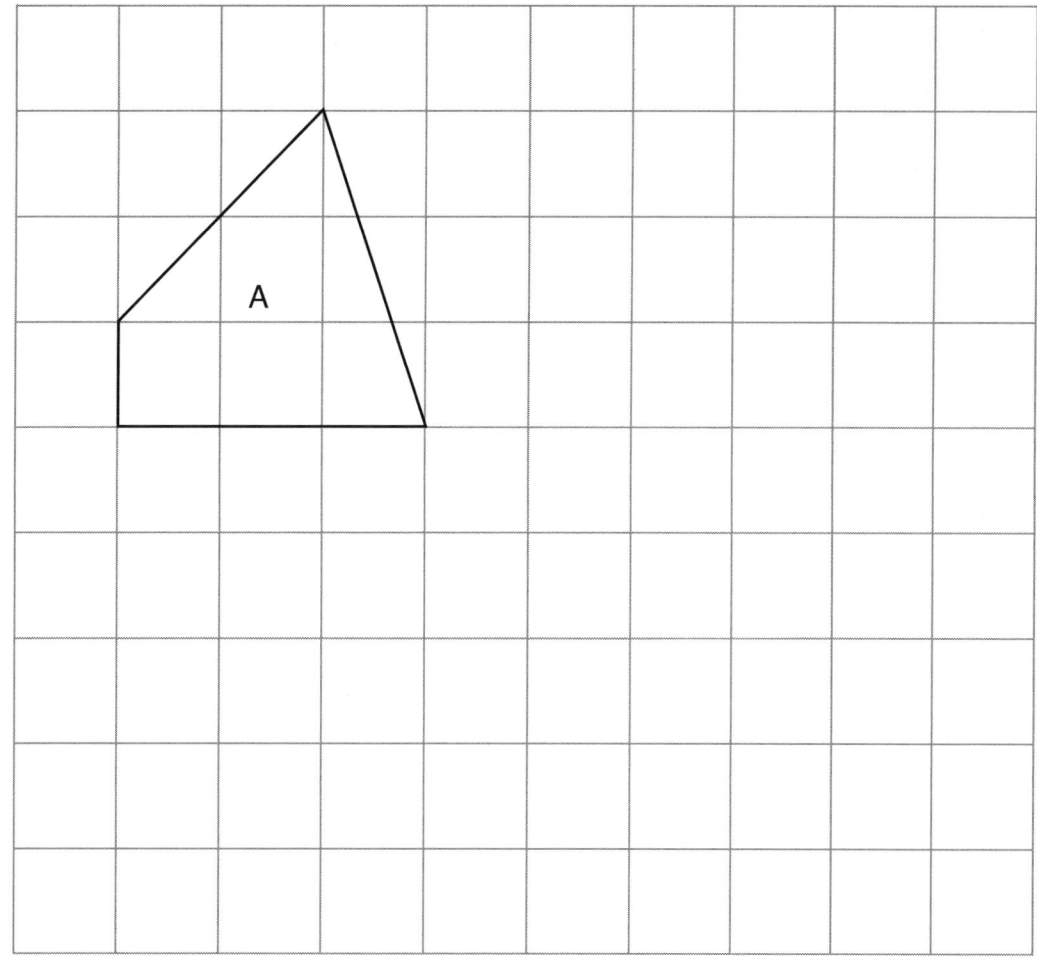

24

2 marks

25 Kim has 4 kg of cement mixture. 20% of this weight is cement and the rest is sand. What weight of sand is in the mixture?

Show your working. You may get a mark.

25

2 marks

End of test

Total out of 4 _____

MATHEMATICS

YEAR 5

TEST 2 PAPER B — LEVELS 3–5

CALCULATOR ALLOWED

PAGE	MARKS
3	
4	
5	
6	
7	
8	
9	
10	
11	
12	
TOTAL	

RESOURCES

- pencil
- ruler
- calculator

Name

Date

Class

Instructions

You may use a calculator to answer any questions in this test.

Work as quickly and as carefully as you can.

You have 45 minutes for this test.

If you cannot do one of the questions, go on to the next one.

You can come back to it later if you have time.

If you have finished before the end, go back and check your work.

Follow the instructions for each question carefully.

 This shows where you need to put your answer.

If you need to do any working out, you can use any space on the page.

Some questions have an answer box like this:

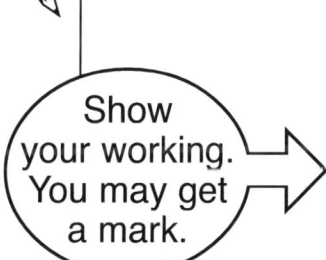

Show your working. You may get a mark.

For these questions you may get a mark for showing your working.

1 Here is a number line.

0   50

Draw an arrow (↓) to show the position of 45.

1 mark

2 Here is a shape on a co-ordinates grid.

Draw the reflection of the shape in the mirror line.

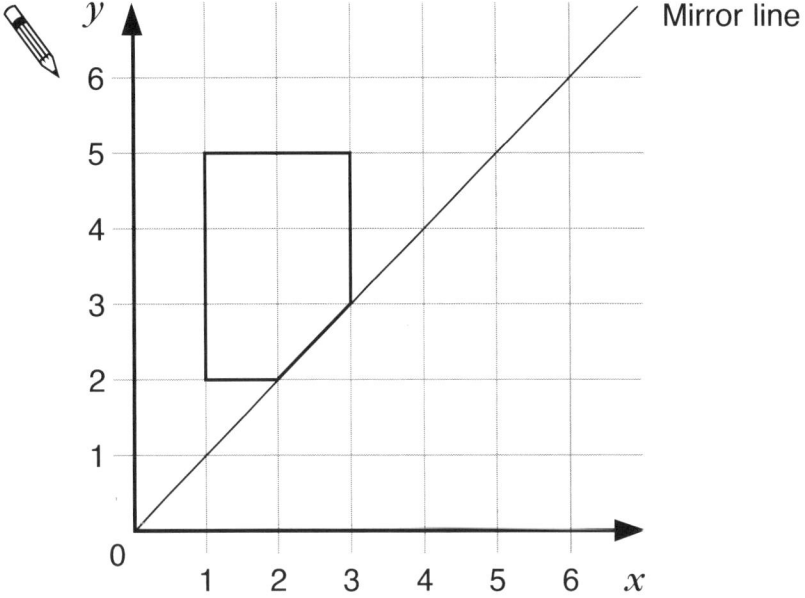

The five corners of the reflected shape you have drawn

have the co-ordinates (2, 2), (3, 3), (____, ____), (____, ____)

and (____, ____).

2

2 marks

3 Calculate 35% of 70.

3

1 mark

4   Sam has some notes and coins.

How much money does Sam have?

£ ⬚

4

---

5   Write a number in each box to make these fractions equivalent.

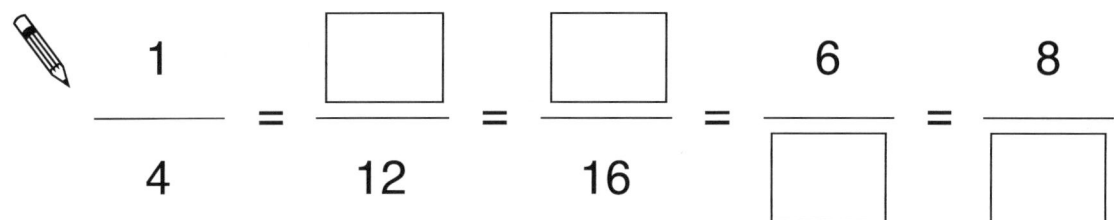

$$\frac{1}{4} = \frac{\boxed{}}{12} = \frac{\boxed{}}{16} = \frac{6}{\boxed{}} = \frac{8}{\boxed{}}$$

5

---

6   Here are some numbers.

4   8   35   3

Write each amount in a box to make this number story correct.

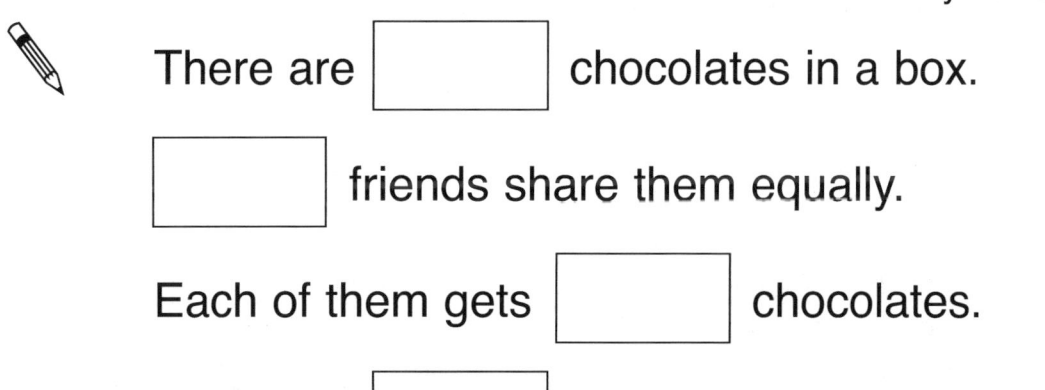

There are ⬚ chocolates in a box.

⬚ friends share them equally.

Each of them gets ⬚ chocolates.

There are ⬚ chocolates left over.

6

---

7 Write all the factors of 24.

7

2 marks

8 The table shows the results of a survey of the heights of some children.

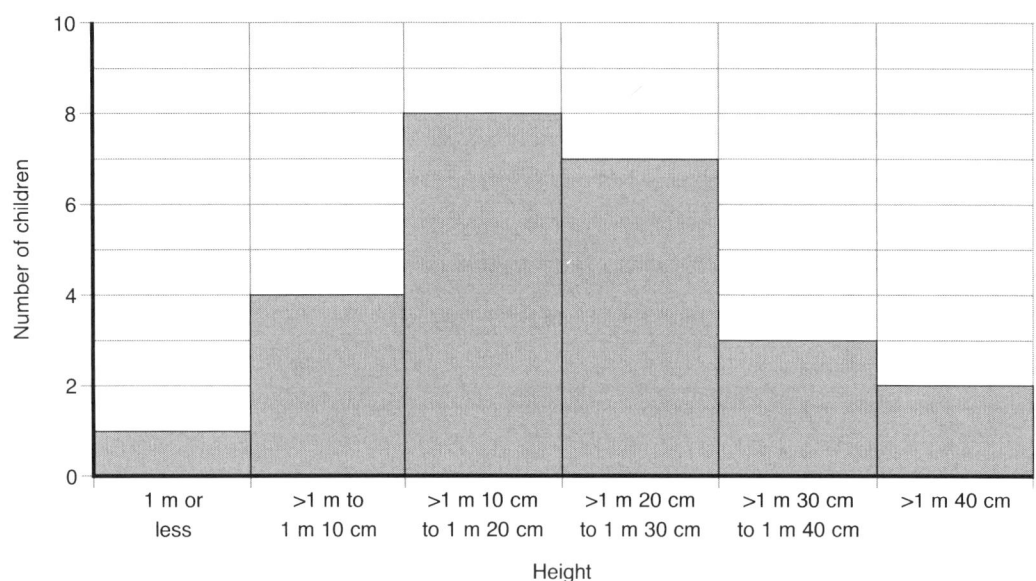

a) How many children are 1 m 30 cm or less?

8a

1 mark

b) How many children were surveyed?

8b

1 mark

9 Write in the missing number.

$$473 \times \boxed{} = 1892$$

9

1 mark

10 $4\frac{1}{4} + 2\frac{3}{4} =$

1 mark

11   Parin is weighing some sugar.

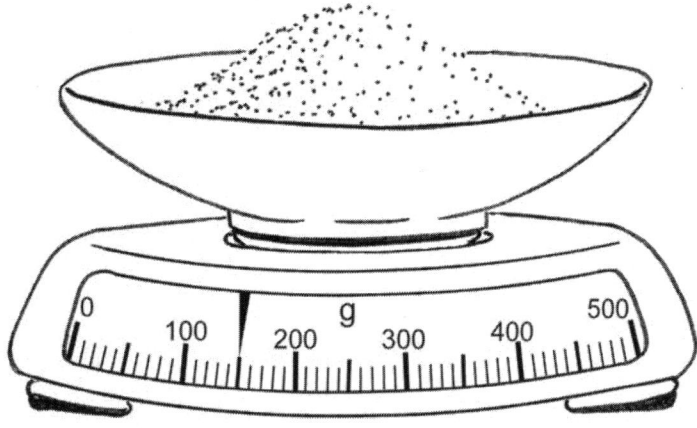

How many more grams of sugar are needed to make 500 g?

11

1 mark

12   What is the area of this shape?

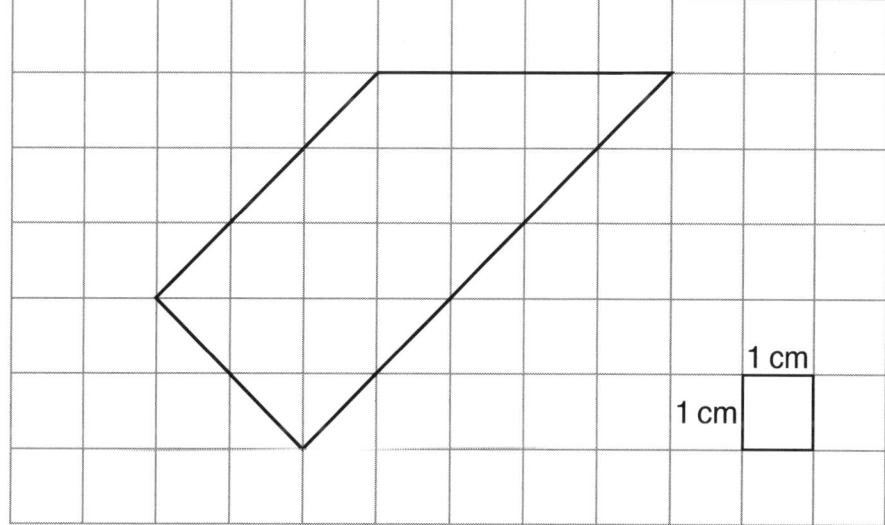

This diagram is not to scale.

12

1 mark

13 Mr. Jenks is cutting his lawn. The lawn mower cuts a line of grass 55 cm wide and the lawn is 17 m wide. What is the minimum number of lines of grass he must cut to mow the whole lawn?

Show your working. You may get a mark.

13

2 marks

14 Write in the missing number.

$$\boxed{} - 53 = 765$$

14

1 mark

15 $b = 12$

Complete the following number sentences.

a) $3 \times b - 10 = \boxed{}$

15a

1 mark

b) $\frac{b}{2} + 4 = \boxed{}$

15b

1 mark

16 Here is a triangle.

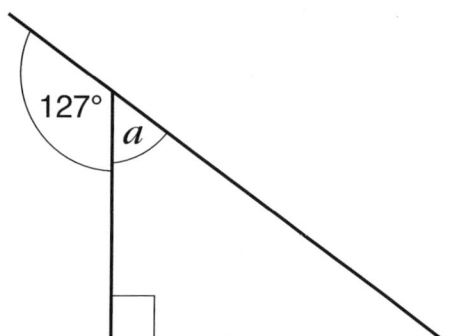

Not to scale

Calculate the size of angle a.

Do not use a protractor (angle measurer).

Show your working. You may get a mark.

16

2 marks

17 Jensen is training for a race. He runs 8 times around a 400 m race track.

a) How far does he run in total?

17a

1 mark

Each circuit of the race track takes him $1\frac{1}{2}$ minutes.

b) How long does he spend running around the race track?

17b

1 mark

18 Sam went to a film starting at 2:15 p.m. The film lasted 1 hour 40 minutes and then he spent $1\frac{1}{2}$ hours shopping.

At what time did he finish shopping?

Show your working. You may get a mark.

18

2 marks

19 Write in the missing number.

$$\frac{76}{\boxed{}} = 0{\cdot}76$$

19

1 mark

20 Complete the table.

	Number of lines of symmetry	Number of corners
semi-circle		
square		
rectangle		
equilateral triangle		
regular hexagon		

20

2 marks

21 The table shows what kind of pets some children have.

	Number of children
Cat	8
Dog	6
Fish	9
Bird	3
None	10

Complete the pictogram.

	Number of children
Cat	☺ ☺ ☺ ☺
Dog	☺ ☺ ☺
Fish	☺ ☺ ☺ ☺ (
Bird	☺ (
None	

Key:

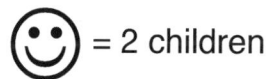 = 2 children

(= 1 child

1 mark

22 Circle the number that is **not** a common multiple of 3 and 4.

24   12   36   42   60

22
1 mark

Total out of 2 _____

23 The table shows the average maximum temperatures over the year in Chicago.

	J	F	M	A	M	J	J	A	S	O	N	D
°C	0	1	6	13	18	24	27	26	23	16	8	2

a) Complete the graph below using this information.

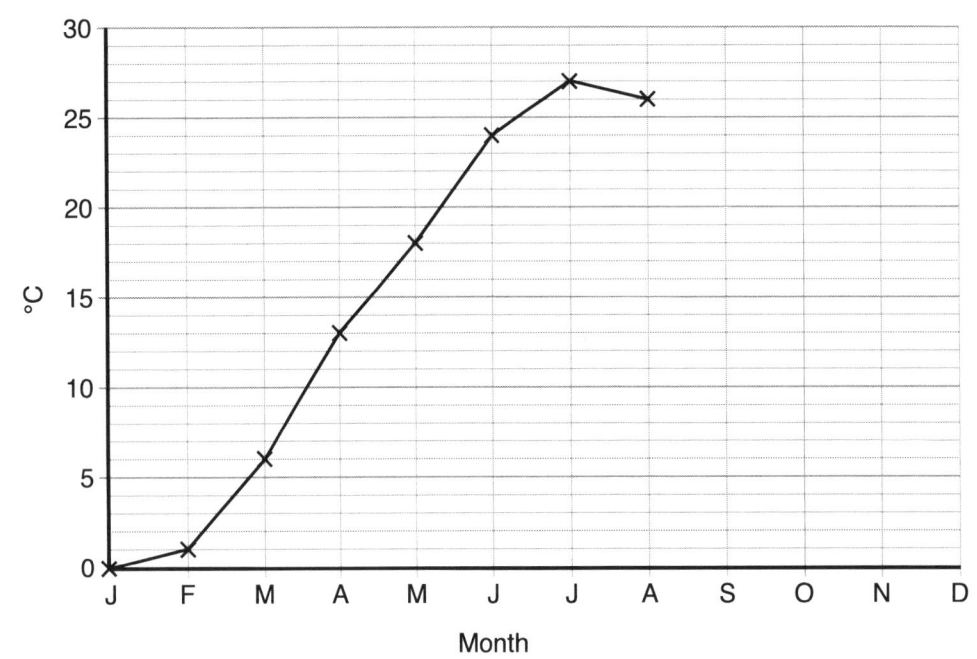

b) Explain why a line graph shows this information better than a bar chart.

23a

1 mark

23b

1 mark

24 Rotate the shape through a quarter of a turn anticlockwise around point A.

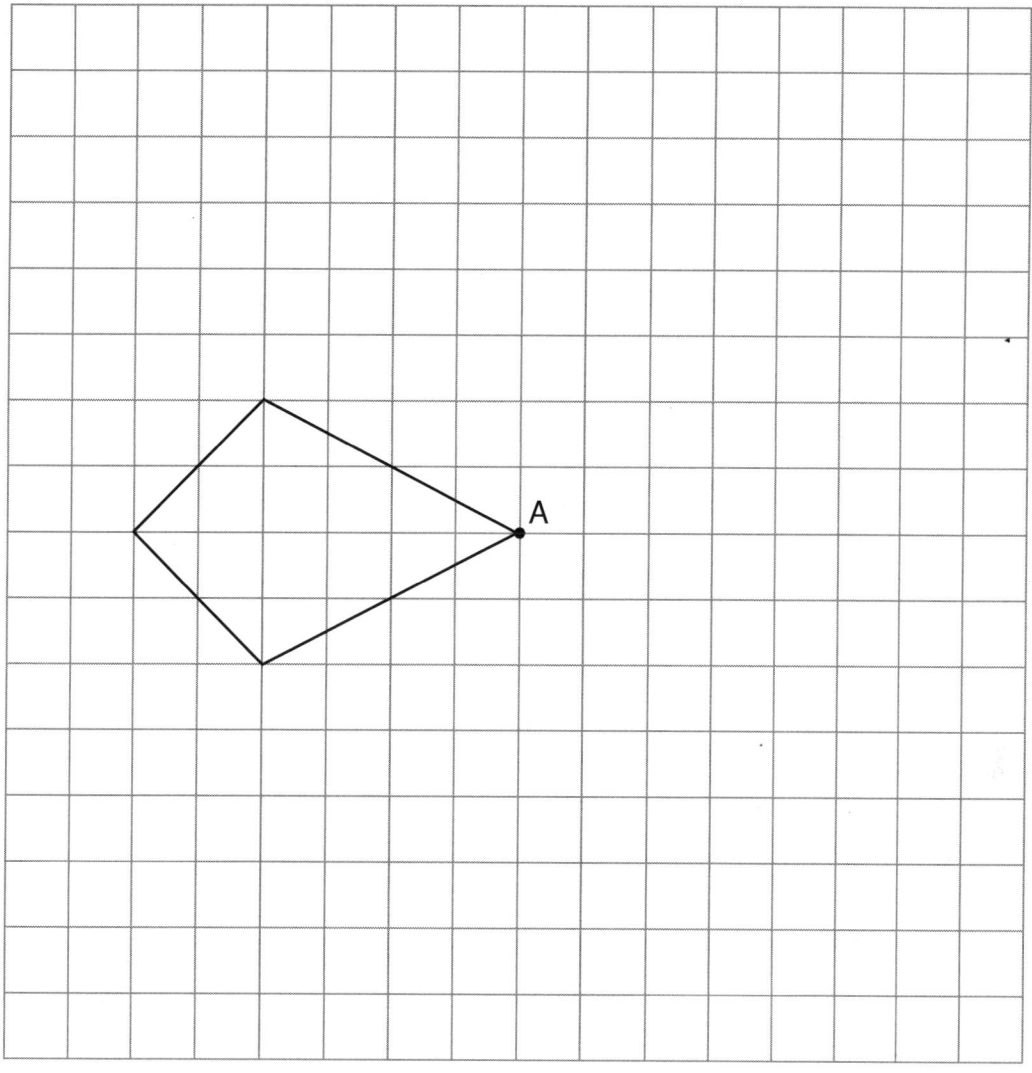

A

End of test

Test 3
Mental mathematics test questions and answers

Say: **For this group of questions, you will have 5 seconds to work out each answer and write it down.**

	The questions	Answers
1.	What is the product of nine and seven?	63
2.	Add together fifty, thirty-five and twenty.	105
3.	What is half of eight point six?	4·3
4.	Subtract two thousand three hundred from six thousand nine hundred.	4600
5.	What is the sum of eight point four and five point six?	14

Say: **For the next group of questions, you will have 10 seconds to work out each answer and write it down.**

		Answers
6.	Look at your answer sheet. Draw a ring around the fraction that is equal to one-fifth.	$\frac{3}{15}$
7.	A packet of Chew Drops cost thirty-five pence each. How much do five packets of Chew Drops cost?	£1.75
8.	Look at your answer sheet. Two hundred millilitres more water is poured into the cylinder. How much water is in the cylinder now?	470 ml
9.	The perimeter of a square is fifty metres. How long is each side?	12·5 m
10.	How many lines of symmetry does a regular hexagon have?	6
11.	What is three-quarters of sixty?	45
12.	What is twenty-five multiplied by eight?	200
13.	Leroy was born in nineteen ninety-eight. In what year will he have his twenty-first birthday?	2019
14.	The temperature is minus two degrees Celsius. It falls by five degrees. What is the new temperature?	–7 °C
15.	I am thinking of a 3-D shape. It has a square base. It has four other faces which are triangles. What is the name of the 3-D shape?	square-based pyramid

Say: **For the next group of questions, you will have 15 seconds to work out each answer and write it down.**

		Answers
16.	Multiply four point three by six.	25·8
17.	What number is double zero point six eight?	1·36
18.	An ice lolly costs ninety pence. How many ice lollies can I buy with ten pounds?	11
19.	What is the difference between one thousand, nine hundred and ninety-three and five thousand and four?	3011
20.	I multiply a number by one hundred. My answer is three hundred and eighty. What was the number I started with?	3·8

Say: **Now put down your pencil. The test is finished.**

Test 3
Papers A and B answers

Paper A

1. 5·731
2. 5   8
3. 79·63   0·796
      1·248   24·698
4. £1.74
5. 4·38   4·5   4·675 $4\frac{7}{10}$ $4\frac{3}{4}$
6. 9
7. 3·15
8.

9.

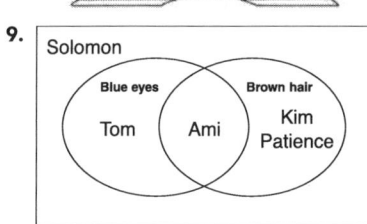

10. 35   67   91
11. £8.20
12. a) 22   b) 116
13. 10
14. 3:10 p.m.
15. 20 cm
16. a) 12   b) £6.79
17. 41 cm²
18. a) 128 litres   b) £144
19. 10 500
20. Accept **any one** of the four shaded squares.

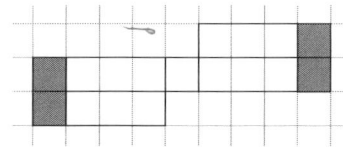

21.

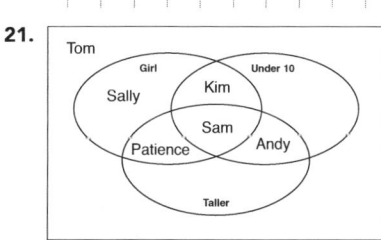

22. a) May   b) 3
23. 1 litre 170 ml (or 1·17 litres or 1170 ml)
24. 1 litre 650 ml (or 1·65 litres or 1650 ml)
25. A 138°   B 48°

Paper B

1. 18
2. $\frac{1}{6} = \frac{2}{12} = \frac{3}{18} = \frac{5}{30} = \frac{6}{36}$
3.

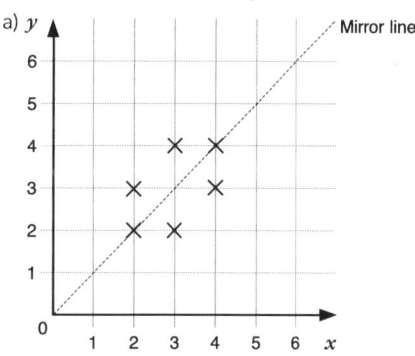

4. £8.40
5. 13·5
6. 200 ml
7. 100
8. 6
9. 32
10. 1, 2, 3, 5, 6, 10, 15, 30
11. 6, 2, 4, 8   or   6, 2, 8, 4
12. 6
13. a) 4   b) 7
14. 10:20 p.m.
15. a) 1 m 14 cm   b) 1 m 28 cm
16. 33 cm²
17. Monday
18. 28°
19.

	No. of edges	No. of vertices
cube	12	8
cuboid	12	8
cylinder	2	0
square-based pyramid	8	5
triangular prism	9	6

20. a) strawberries   b) grapes
21. a) 500 ml (or $\frac{1}{2}$ *l*)   b) 8 days
22. a)

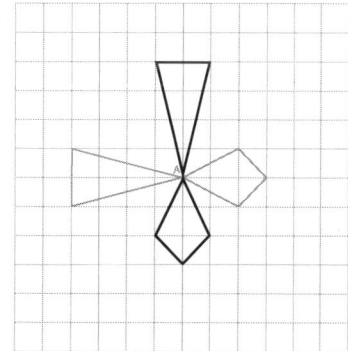

    b) (3, 2) and (4, 3)
23. 14
24.

Mental mathematics test

Name ....

Date .... Class ....

Total marks

Time: 5 seconds

1		

2		50 35 20

3		8·6

4		2300 6900

5		8·4 5·6

Time: 10 seconds

6	$\frac{2}{10}$ $\frac{3}{10}$ $\frac{3}{15}$ $\frac{2}{15}$ $\frac{4}{25}$

7	£	35p

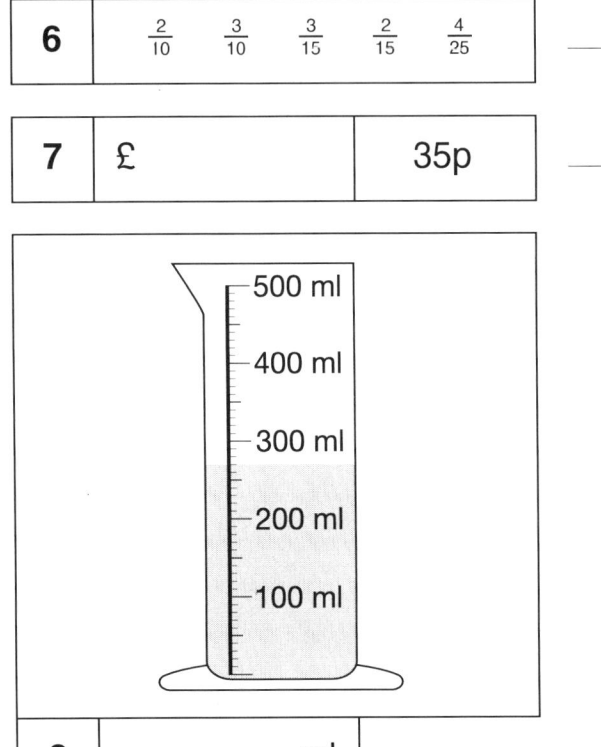

8		ml

9	m	50 m

10		

11		60

12		25 8

13		1998

14	°C	−2 °C

15		

Time: 15 seconds

16		4·3 6

17		0·68

18		90p £10

19		1993 5004

20		380

Year 5 Test 3

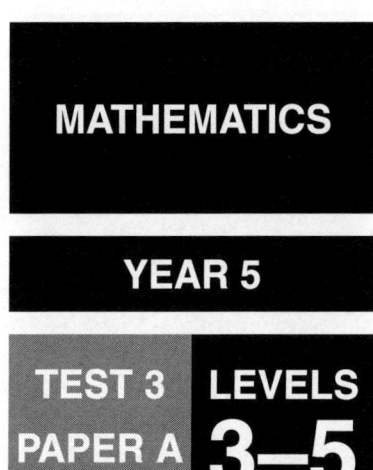

MATHEMATICS

YEAR 5

TEST 3 PAPER A LEVELS 3–5

CALCULATOR
NOT ALLOWED

PAGE	MARKS
3	
4	
5	
6	
7	
8	
9	
10	
11	
12	
TOTAL	

RESOURCES
- pencil
- ruler

Name

Date

Class

Instructions

You may not use a calculator to answer any questions in this test.

Work as quickly and as carefully as you can.

You have 45 minutes for this test.

If you cannot do one of the questions, go on to the next one.

You can come back to it later if you have time.

If you have finished before the end, go back and check your work.

Follow the instructions for each question carefully.

 This shows where you need to put your answer.

If you need to do any working out, you can use any space on the page.

Some questions have an answer box like this:

Show your working. You may get a mark.

For these questions you may get a mark for showing your working.

1 Circle the decimal closest to 6.

 5·489   7·366   6·812   5·731   6·279

1 mark

1

2 Round the following decimals to the nearest whole number.

 4·8    8·24

1 mark

2

3 Divide each of the following numbers by 100.

 7963    79·6

 124·8    2469·8

1 mark

3

4 Calculate £3.50 − £1.76

1 mark

4

5 Order the following fractions and decimals, smallest to largest.

4·675 $4\frac{3}{4}$ 4·38   4·5 $4\frac{7}{10}$

smallest

1 mark

5

6 Write in the missing digit.

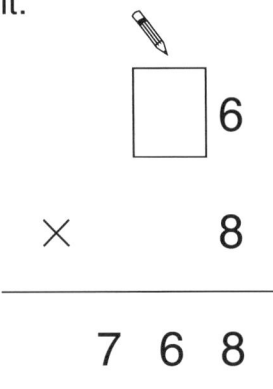

$$\begin{array}{r} \boxed{}\,6 \\ \times \qquad 8 \\ \hline 7\ 6\ 8 \\ \hline \end{array}$$

6

1 mark

7 Here is a number line.

Write the number shown by the arrow.

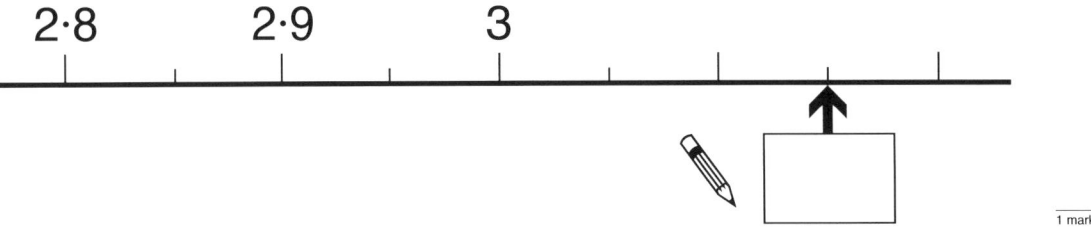

2·8 2·9 3

7

1 mark

8 Draw the hands on the clock to show the time of 19:35.

8

1 mark

9 5 children recorded whether their eyes were blue
and their hair was brown.

	Blue eyes	Brown hair
Tom	✓	
Kim		✓
Solomon		
Ami	✓	✓
Patience		✓

Write each child's name in the correct region of the
Venn diagram. One has been done for you.

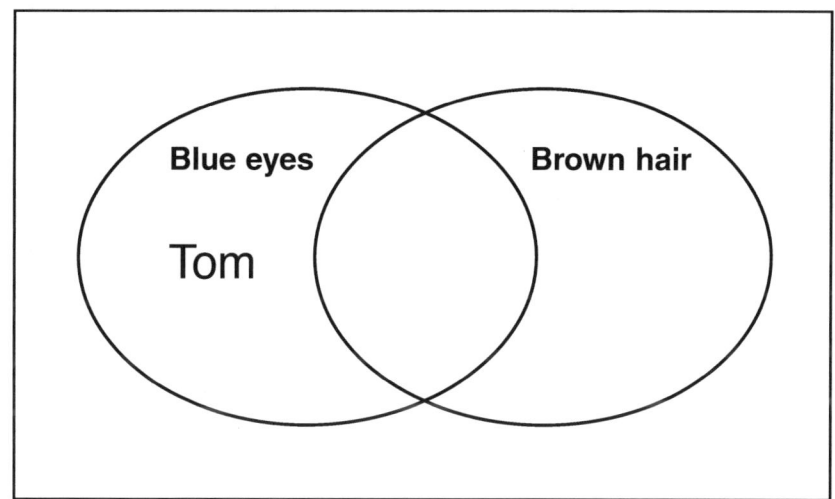

9

___
2 marks

10 Circle all the numbers that are 3 more than a multiple of 8.

17   35   49   67   78   91

10
___
1 mark

11 Sam buys 4 bars of soap. They cost 45p each.

How much change does she get from a £10 note?

Show your working. You may get a mark.

11

2 marks

12 The table shows the number of CDs and DVDs for sale in a shop.

	CDs	DVDs
less than £10	67	22
£10 or more	49	53

a) How many DVDs cost under £10?

12a

1 mark

b) How many CDs are for sale?

12b

1 mark

13 Salma has 2·5 litres of squash.

How many 250 ml glasses can she fill?

13

1 mark

14 Tina arrives home at 5:30 p.m. after a journey
of 2 hours 20 minutes.
At what time did she start her journey?

1 mark   14

15 This shape is made up of 1 centimetre squares.
What is its perimeter?

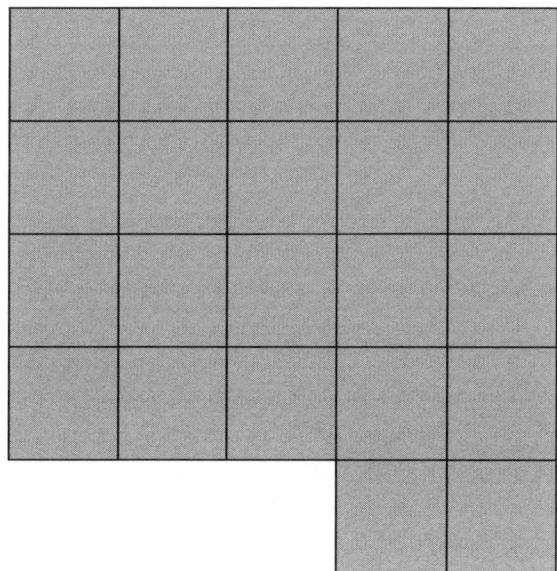

This diagram
is not to scale.

1 mark   15

16 At the newsagents, Kim spends £5.40 on 45p stamps and
£1.39 on a magazine.

a) How many stamps did Kim buy?

1 mark   16a

b) How much did she spend altogether?

1 mark   16b

17 What is the area of this shape?

This diagram is not to scale.

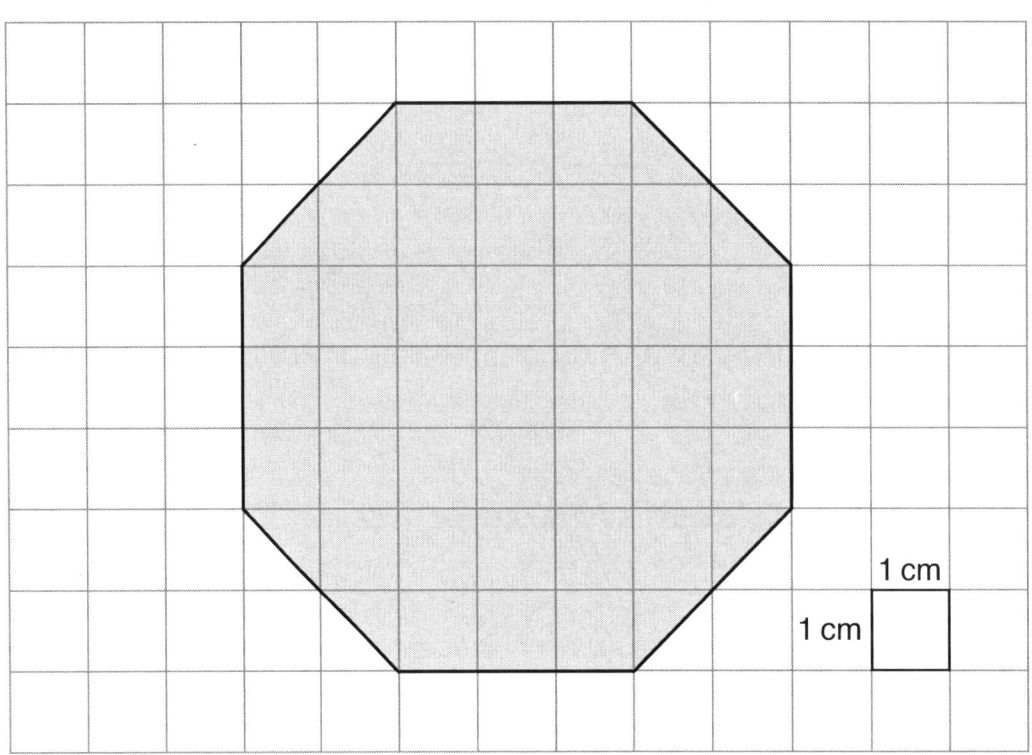

1 cm

1 cm

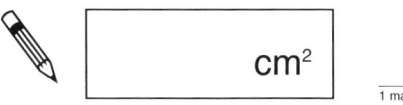
cm²

1 mark

18 Connor uses 32 litres of petrol in his motorbike each week.

a) How many litres does he use in four weeks?

18a
1 mark

b) He spends £36 each week on petrol. How much does he spend in four weeks?

18b
1 mark

19 Joe the delivery man has 15 boxes of biscuits to deliver. Each box has 28 packets of biscuits in it, with 25 biscuits in each packet. How many biscuits does he have to deliver in total?

Show your working. You may get a mark.

19

2 marks

20 Complete the net of this cuboid.

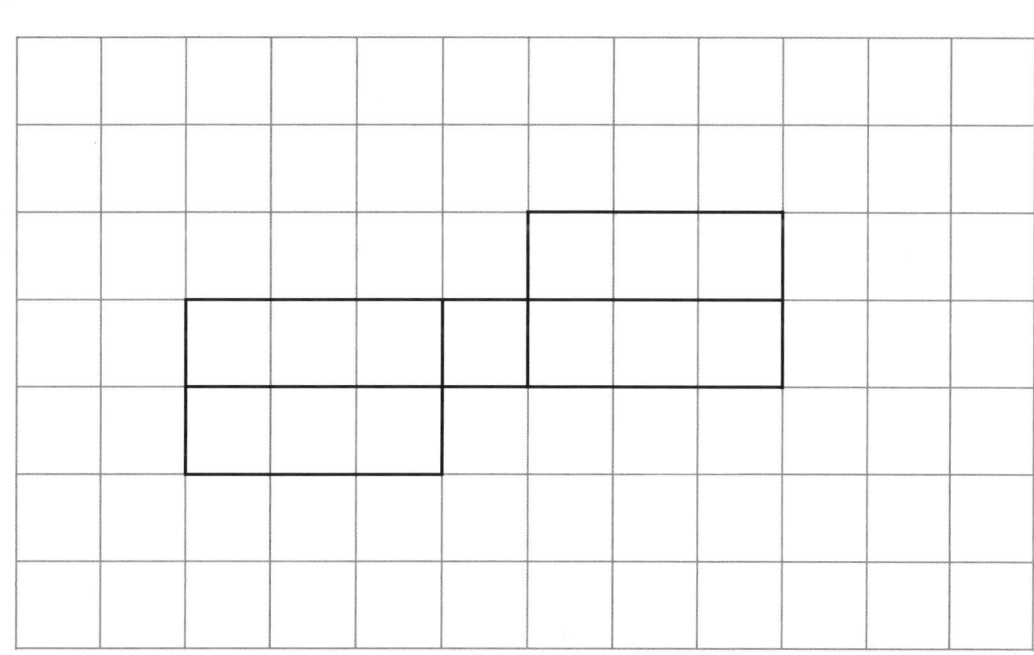

20

1 mark

Cindy has done a survey of six of her friends.

	Girl	Under 10 years	Taller than Cindy
Tom			
Kim	✓	✓	
Sam	✓	✓	✓
Sally	✓		
Andy		✓	✓
Patience	✓		✓

Write these friends' names in the correct place on the Venn diagram.

One has been done for you.

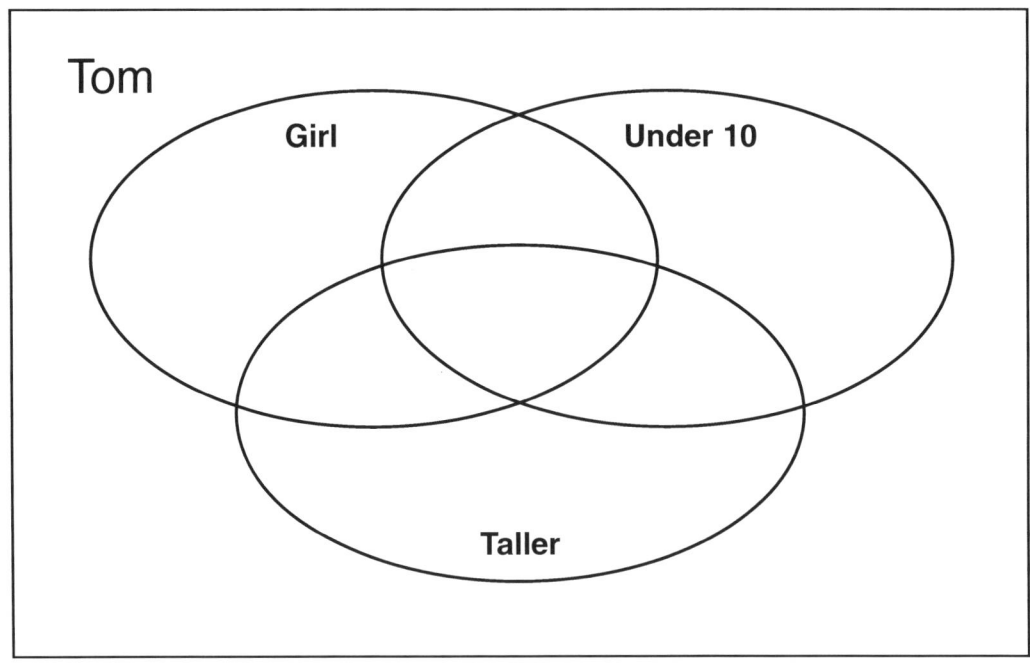

21

2 marks

22 This graph shows the maximum and minimum temperatures in Moscow during the year.

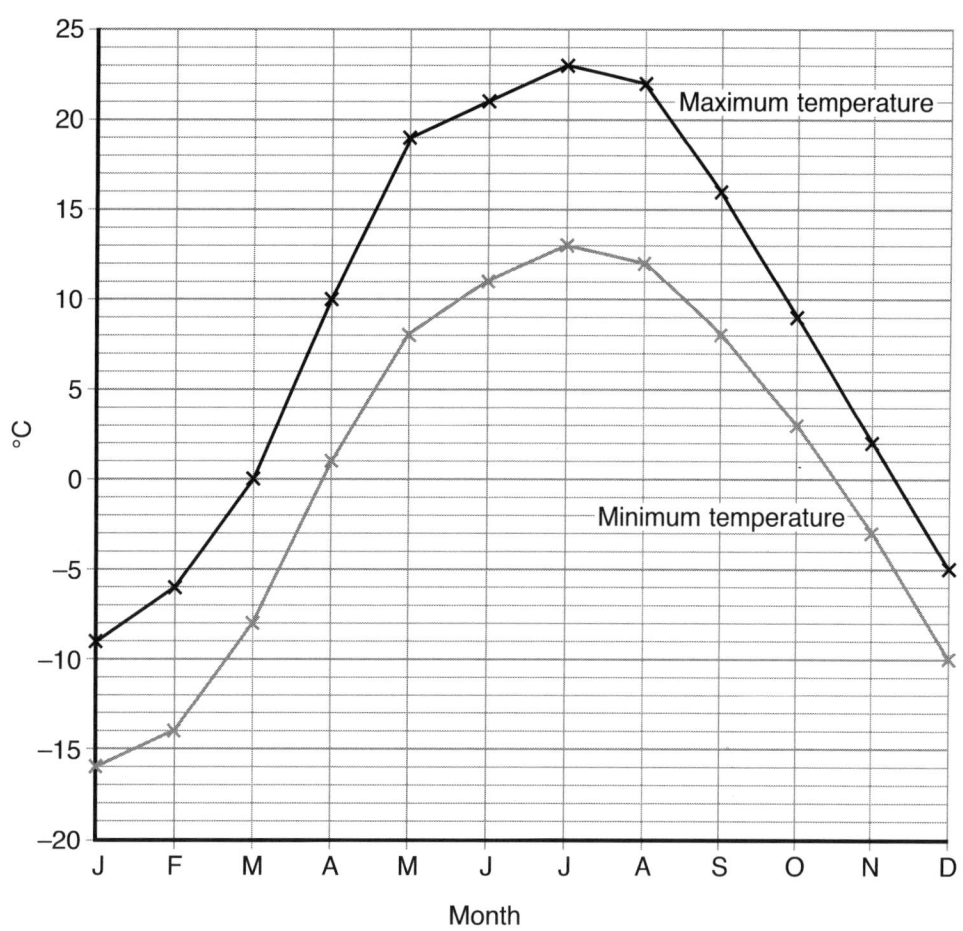

a) In which month is the difference between the maximum and minimum temperatures the greatest?

22a

1 mark

b) For how many months of the year is the minimum temperature above 8 °C?

22b

1 mark

23 John has 1·8 litres of milk. He uses 65% of this making custard. How much milk does he use?

Show your working. You may get a mark.

23

2 marks

24 How much water is in this jug?

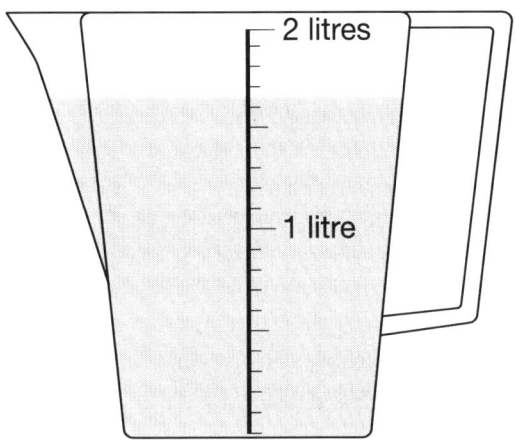

- 2 litres

- 1 litre

24

1 mark

25 Look at the triangle below.

Calculate the size of angles A and B.

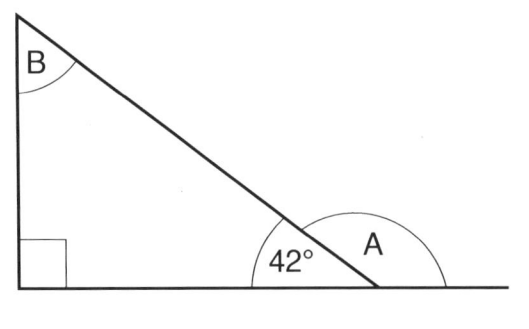

This diagram is not to scale.

B

42°   A

A   B

25

2 marks

End of test

Total out of 5 _____

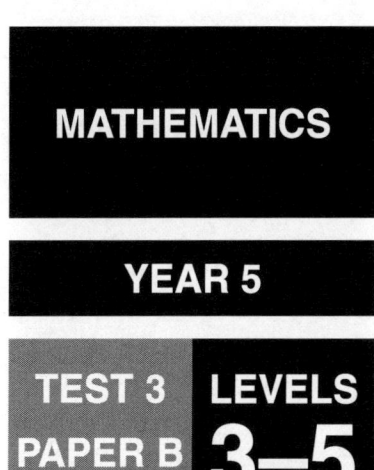

MATHEMATICS

YEAR 5

TEST 3 PAPER B

LEVELS 3–5

CALCULATOR ALLOWED

PAGE	MARKS
3	
4	
5	
6	
7	
8	
9	
10	
11	
12	
TOTAL	

RESOURCES

- pencil
- ruler
- calculator

Name

Date

Class

Instructions

You may use a calculator to answer any questions in this test.

Work as quickly and as carefully as you can.

You have 45 minutes for this test.

If you cannot do one of the questions, go on to the next one.

You can come back to it later if you have time.

If you have finished before the end, go back and check your work.

Follow the instructions for each question carefully.

 This shows where you need to put your answer.

If you need to do any working out, you can use any space on the page.

Some questions have an answer box like this:

For these questions you may get a mark for showing your working.

1 Circle the number that is **not** a common multiple of 6 and 4?

60   24   36   72   18

1

1 mark

2 Write a number in each box to make these fractions equivalent.

$$\frac{1}{6} = \frac{\boxed{}}{12} = \frac{3}{\boxed{}} = \frac{\boxed{}}{30} = \frac{6}{\boxed{}}$$

2

2 marks

3 Here is a number line.

0   100

Draw an arrow (↓) to show the position of 75.

3

1 mark

4 Colin has a note and some coins.

How much money does Colin have?

£ _____

4

1 mark

5   Calculate 15% of 90.

1 mark

6   Here is a jug with some milk in it.

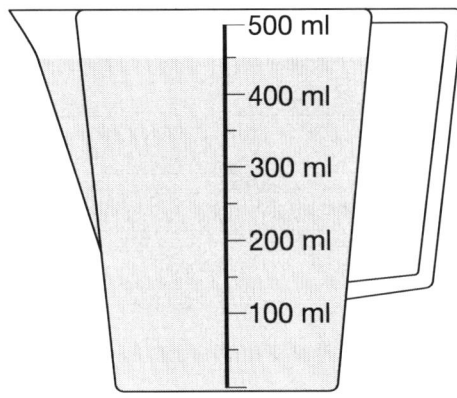

Simone uses 250 ml of milk to make pancakes.

How many millilitres of milk are left in the jug?

6
1 mark

7   Write in the missing number.

$$5 \cdot 2 \times \boxed{} = 520$$

7
1 mark

8 $3\frac{2}{5} + 2\frac{3}{5} = \boxed{}$

8
1 mark

9 Write in the missing number.

$$\boxed{} + 587 = 619$$

1 mark

10 Write all the factors of 30.

10

2 marks

11 Here are some numbers.

$$4 \quad 8 \quad 2 \quad 6$$

Write each amount in a box to make this number story correct.

There are $\boxed{}$ eggs in a carton.

Louise buys $\boxed{}$ cartons.

She uses $\boxed{}$ to make a cake.

There are $\boxed{}$ eggs left.

11

1 mark

Total out of 4 _____

12 Write in the missing number.

$$796 \times \boxed{} = 4776$$

12

1 mark

13 d is the square of the number a.

Complete the following sentences.

a) If $d = 16$, then $a = \boxed{}$

13a

1 mark

b) If $d = 49$, then $a = \boxed{}$

13b

1 mark

14 Danni went to a play starting at 7:30 p.m. The first act lasted 1 hour 10 minutes and then there was a 20 minute interval. The second act lasted 1 hour 20 minutes.

At what time did the second act finish?

Show your working. You may get a mark.

14

2 marks

15 Cindy is 1 m 42 cm tall. Ami is 28 cm shorter than Cindy.

Ellie is exactly halfway between them in height.

a) How tall is Ami?

15a

1 mark

b) How tall is Ellie?

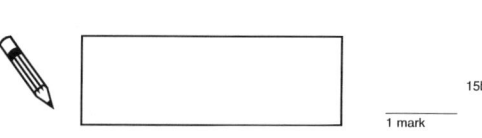

15b

1 mark

16 What is the area of this shape?

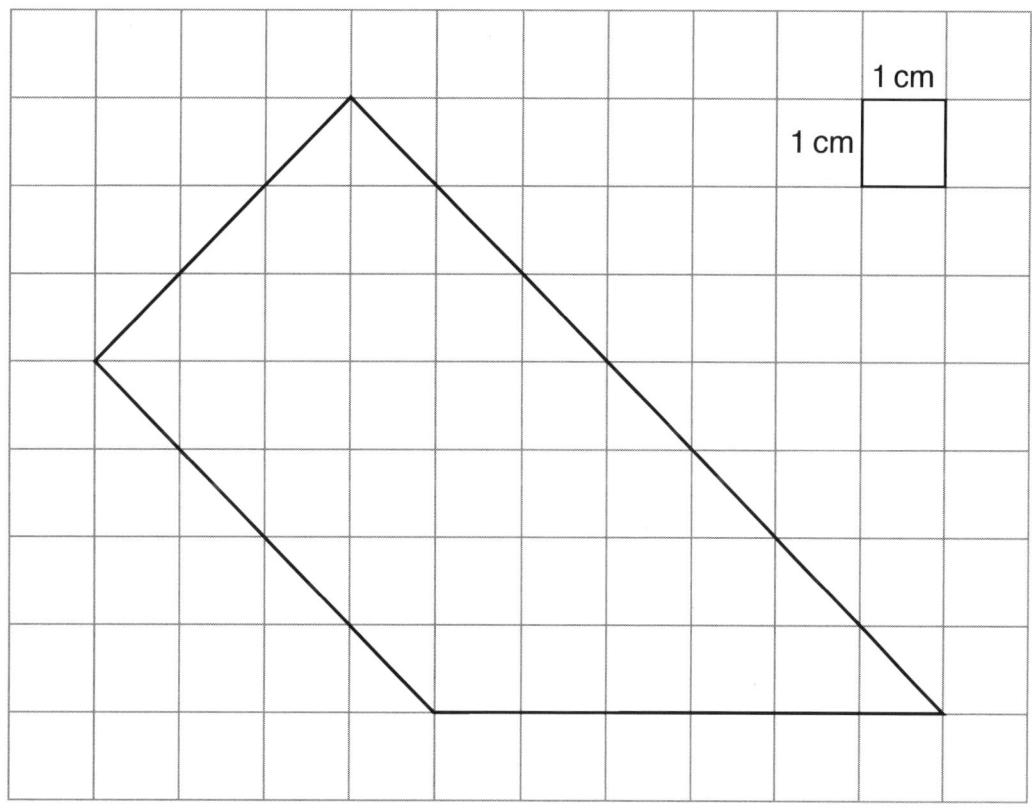

This diagram is not to scale.

16

1 mark

17 Here is a calendar for March.

March					
Sunday		5	12	19	26
Monday		6	13	20	27
Tuesday		7	14	21	28
Wednesday	1	8	15	22	29
Thursday	2	9	16	23	30
Friday	3	10	17	24	31
Saturday	4	11	18	25	

What day of the week is the 10th July?

Show your working. You may get a mark.

17

2 marks

18 Here is a triangle.

Not to scale

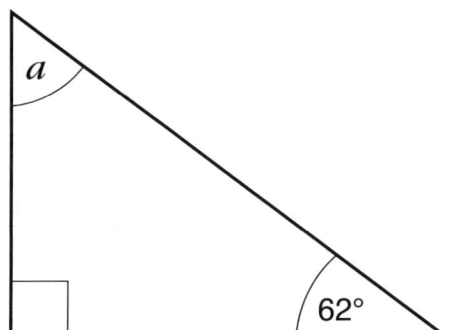

Calculate the size of angle a.

Do not use a protractor (angle measurer).

Show your working. You may get a mark.

18

2 marks

19 Complete the table.

	Number of edges	Number of vertices
cube		
cuboid		
cylinder		
square-based pyramid		
triangular prism		

19

2 marks

20   The table shows the favourite fruit of some Year 5 children.

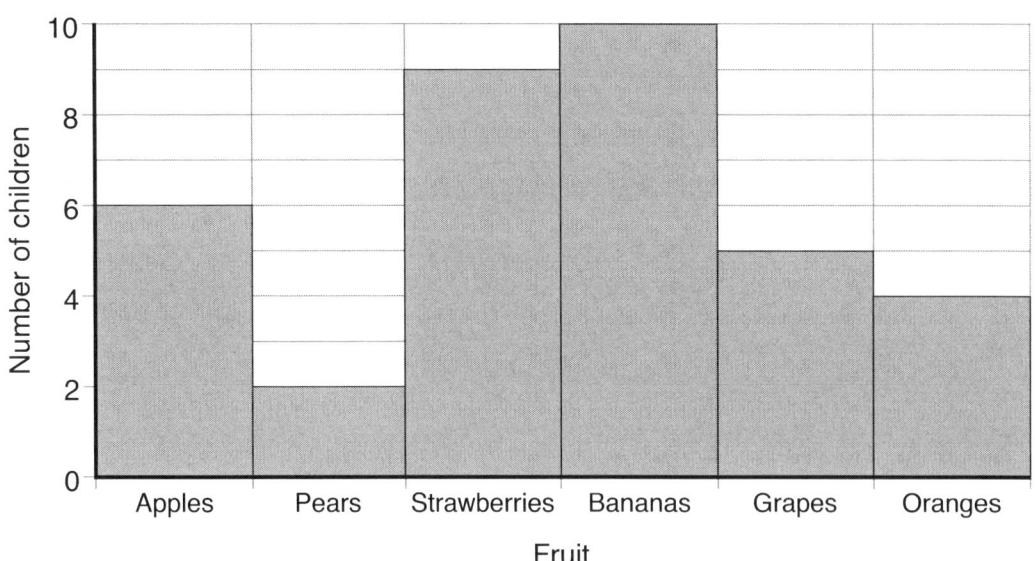

a) Which fruit did 3 more children prefer to apples?

20a

1 mark

b) Which fruit was half as popular as bananas?

20b

1 mark

21   Rashid has 4 litres of milk. He uses 12·5% of it in one day.

a) How much milk does he use in one day?

21a

1 mark

b) How many days will his milk last him if he
uses the same amount each day?

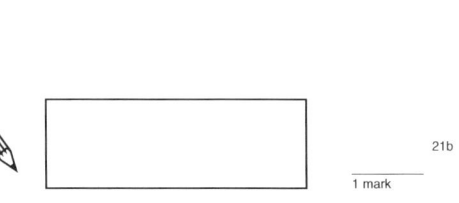

21b

1 mark

22 Mark the points (2, 2), (2, 3), (3, 4) and (4, 4) on the co-ordinates grid.

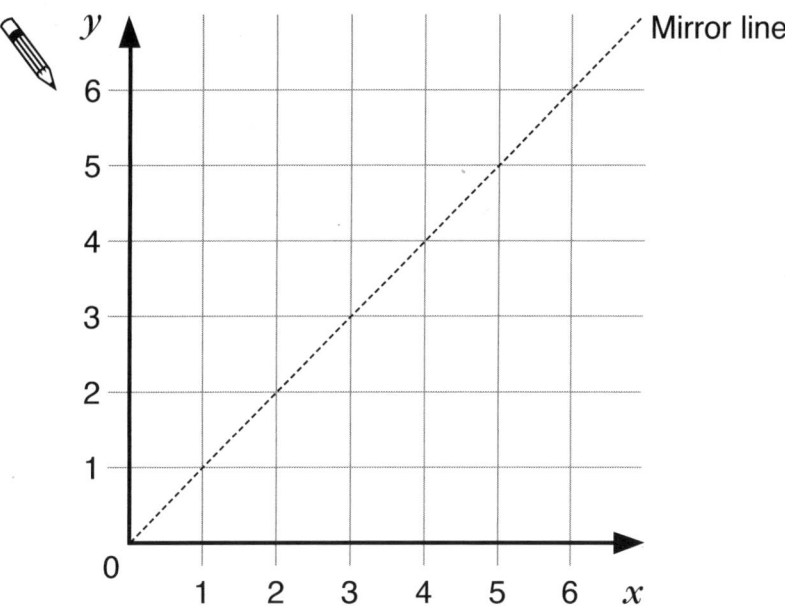

22a

1 mark

Reflect the shape formed about the mirror line to form a hexagon.

The co-ordinates of the two new points are:

 (____ , ____) and (____ , ____)

22b

1 mark

23 Look at this Venn diagram.

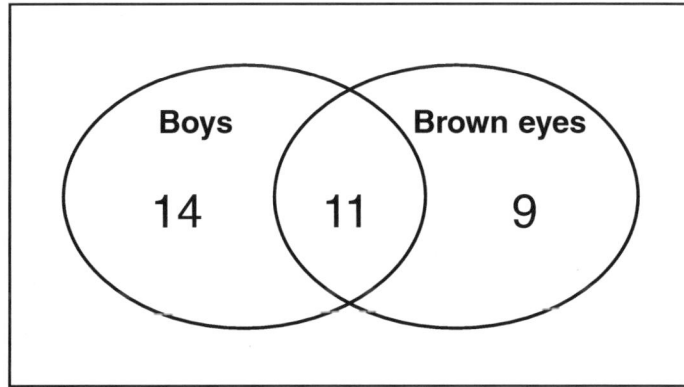

How many boys do not have brown eyes?

23

1 mark

24 Rotate this shape a quarter of a turn clockwise around point A.

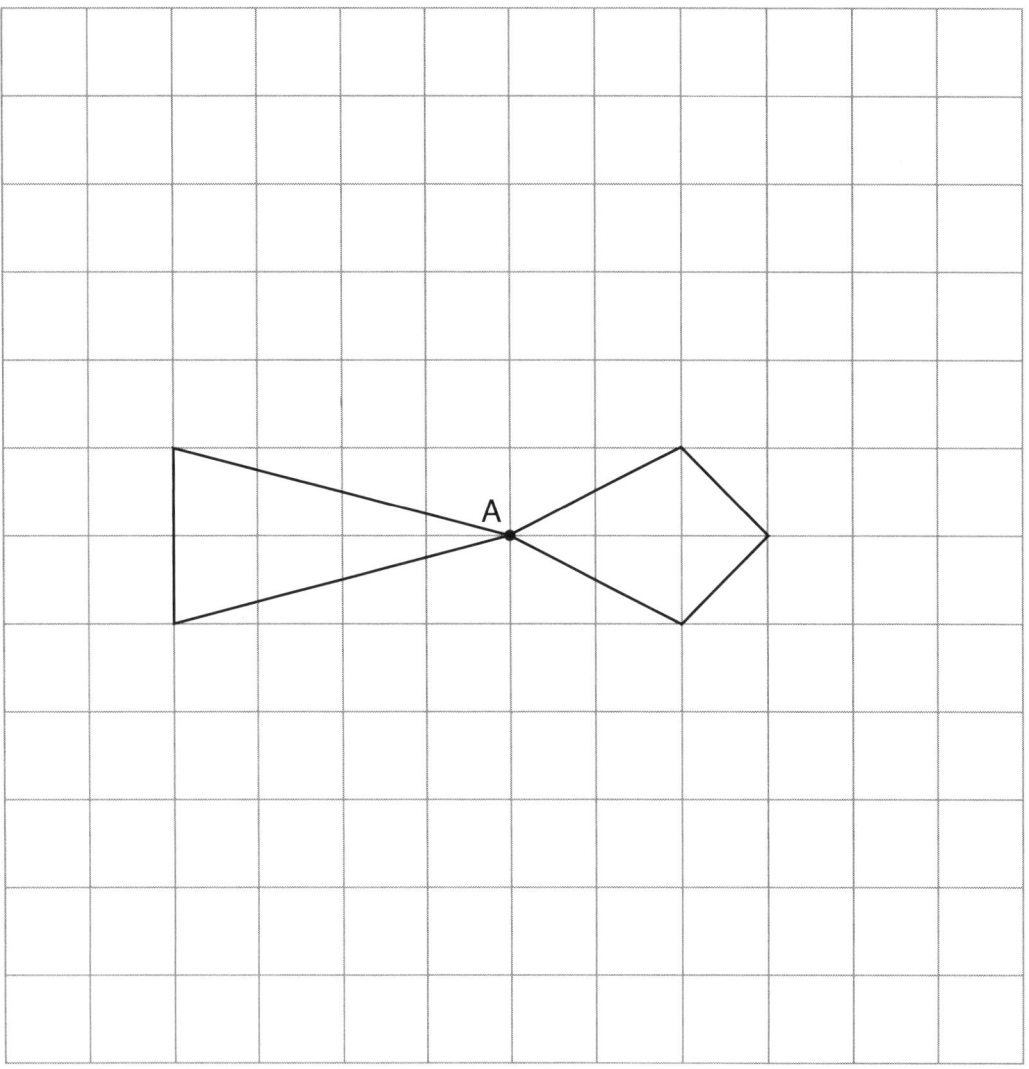

End of test

Record-keeping format 1 Adult Directed Task assessment sheet

Objective(s): _____   Date: _____   Adult: _____

_____   NC Level: _____   Class: _____

Child's name	Success criteria				Other observations	Objective(s) achieved	Future action

Record-keeping format 2 Test 1 grid for test analysis (Mental mathematics test)

Topic	AT	Question	Mark
Multiplication tables	2	1	1
Addition: more than two numbers	2	2	1
Subtraction	2	3	1
Halving whole numbers	2	4	1
Addition: decimals	2	5	1
Addition: money	1 & 2	6	1
Time differences	1 & 3	7	1
Fraction of amounts	2	8	1
Distance difference	3	9	1
Temperature difference	2	10	1
Multiplication: money	1 & 2	11	1
Fraction and decimal equivalences	2	12	1
Subtraction	2	13	1
Perimeter	3	14	1
2-D shapes/symmetry	3	15	1
Division: money	1 & 2	16	1
Difference: near multiples of 100	2	17	1
Multiplying/dividing whole numbers by 10	2	18	1
Subtraction: decimals	2	19	1
Doubling whole numbers	2	20	1

Name: 1. 2. 3. 4. 5. 6. 7. 8. 9. 10. 11. 12. 13. 14. 15. 16. 17. 18. 19. 20. 21. 22. 23. 24. 25. 26. 27. 28. 29. 30.

Number correct
Number incorrect or omitted
Percentage correct
Percentage incorrect or omitted

Mental mathematics test score (out of 20)

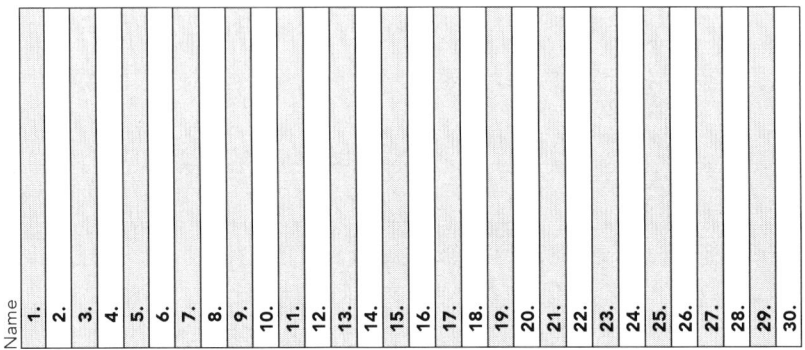

Record-keeping format 2 Test 1 grid for test analysis (Paper A)

Topic	AT	Q	Marks
Ordering and rounding decimals	2	1	1
Multiplication (inverse)	2	2	1
a) Multiplication: money b) Subtraction: money	1 & 2	3ab	2
Multiplying whole numbers and decimals by 100	2	4	1
Ordering decimals	2	5	1
Fraction and percentage equivalences	2	6	1
Multiples	2	7	1
Whole number place value	2	8	1
Addition: money	2	9	1
Handling data: Carroll diagram	4	10ab	2
Time: 24-hour notation	3	11	1
Time difference	1, 2 & 3	12	1
a) Multiplication: money b) Division: money	1 & 2	13ab	2
Perimeter	3	14	1
Handling data: bar chart	4	15ab	2
Length: doubling	1, 2 & 3	16	1
Area	3	17	1
a) Multiplication: length b) Multiplication: money	1, 2 & 3	18ab	2
Multiplication	1 & 2	19	2
3-D shape: nets	3	20	1
Handling data: line graph	4	21ab	2
Length/distance: reading and interpreting scales	3	22	1
Handling data: probability	1 & 4	23	2
Area	3	24	2
Percentages of amounts: money	1 & 2	25	2

Pupils: 1. 2. 3. 4. 5. 6. 7. 8. 9. 10. 11. 12. 13. 14. 15. 16. 17. 18. 19. 20. 21. 22. 23. 24. 25. 26. 27. 28. 29. 30.

169

Record-keeping format 2   Test 1 grid for test analysis (Paper B)

Question / Topic	NC Level	Q no.	Marks
Addition (inverse)	2	1	1
Addition: money	2	2	1
Area	3	3	1
Dividing whole number by 100	2	4	1
Ordering whole numbers	2	5	1
Common multiples	2	6	1
Length/distance	3	7	1
Equivalent fractions	2	8	2
Co-ordinates	3	9ab	2
Percentages of numbers	2	10	1
Understanding word problems	1 & 2	11	1
Time difference	1 & 2	12	2
Multiplication (inverse)	2	13	1
Handling data: table	4	14ab	2
Length: multiplication/addition	1, 2 & 3	15	2
Angles in a triangle	3	16	2
Mass: addition and subtraction	1, 2 & 3	17	2
Algebra	1 & 2	18ab	2
Time: calendar	3	19	1
Addition: fractions	2	20	1
2-D shape: lines of symmetry	3	21	2
Percentages	1 & 2	22	2
Factors	2	23	2
Reflection	3	24	1
Paper B score (out of 35)			
Total score (out of 90)			
National Curriculum Level			

Pupil rows: 1. to 30. (blank)

Record-keeping format 3 Test 2 grid for test analysis (Mental mathematics test)

Topic	AT	Question	Mark
Doubling decimals	2	1	1
Division tables	2	2	1
Addition: decimals	2	3	1
Difference	2	4	1
Multiplying multiples of 10	2	5	1
Multiplication: money	1 & 2	6	1
Fraction of amounts	2	7	1
Addition: time	1 & 3	8	1
Fraction and decimal equivalents	2	9	1
3-D shape	3	10	1
Area	3	11	1
Subtraction	2	12	1
Temperature difference	2	13	1
Mass: scales	3	14	1
Perimeter	3	15	1
Addition: decimals	2	16	1
Halving whole numbers	2	17	1
Addition	2	18	1
Multiplying/dividing whole numbers by 1000	2	19	1
Division: money	1 & 2	20	1

Name: 1. 2. 3. 4. 5. 6. 7. 8. 9. 10. 11. 12. 13. 14. 15. 16. 17. 18. 19. 20. 21. 22. 23. 24. 25. 26. 27. 28. 29. 30.

Mental mathematics test score (out of 20)

Number correct

Number incorrect or omitted

Percentage correct

Percentage incorrect or omitted

Record-keeping format 3   Test 2 grid for test analysis (Paper A)

Topic	Strand	Q	Marks
Ordering decimals	2	1	1
Decimal and fraction equivalences	2	2	1
Multiples	2	3	1
Rounding whole numbers	2	4	1
Division (inverse)	2	5	1
Dividing whole numbers and decimals by 10	2	6	1
Handling data: table	4	7ab	2
Time: 24-hour notation	3	8	1
a) Multiplication: money b) Subtraction: money	1 & 2	9ab	2
Ordering and rounding decimals	2	10	1
Subtraction: money	2	11	1
Handling data: Venn diagram	4	12ab	2
Time difference	1, 2 & 3	13	1
Division: mass	1, 2 & 3	14	1
Perimeter	3	15	1
Subtraction and division: money	1 & 2	16	2
Area	3	17	1
a) Multiplication: mass b) Multiplication: money	1, 2 & 3	18ab	2
Multiplication	1 & 2	19	2
3-D shape: nets	3	20	1
Handling data: line graph	4	21ab	2
Mass: reading and interpreting scales	3	22	1
Handling data: bar chart	4	23	2
Translation	3	24	2
Percentages of amounts: mass	1, 2 & 3	25	2

Pupil rows: 1. – 30. (blank)

Paper A score (out of 35)

Record-keeping format 3 Test 2 grid for test analysis (Paper B)

Topic	NC Level	Question	Marks
Ordering whole numbers	2	1	1
Co-ordinates	3	2	2
Percentages of numbers	2	3	1
Addition: money	2	4	1
Equivalent fractions	2	5	2
Understanding word problems	1 & 2	6	1
Factors	2	7	2
Handling data: bar chart	4	8ab	2
Multiplication (inverse)	2	9	1
Addition: fractions	2	10	1
Mass: calculating with scales	1,2 & 3	11	1
Area	3	12	1
Rounding after division: length	1,2 & 3	13	2
Subtraction (inverse)	2	14	1
Algebra	1 & 2	15ab	2
Angles in a straight line and in a triangle	3	16	2
a) Multiplication: length b) Multiplication: time	1,2 & 3	17ab	2
Calculating: time	1,2 & 3	18	2
Dividing whole numbers by 100	2	19	1
Properties of 2-D shapes	3	20	2
Handling data: pictogram	4	21	1
Common multiples	2	22	1
Handling data: line graph	4	23ab	2
Rotation	3	24	1

Student rows: 1. to 30.

Additional columns:
- Paper B score (out of 35)
- Total score (out of 90)
- National Curriculum Level

Record-keeping format 4 Test 3 grid for test analysis (Mental mathematics test)

Category	AT	Question	Mark
Multiplication tables	2	1	1
Addition: more than two numbers	2	2	1
Halving decimals	2	3	1
Subtraction	2	4	1
Addition: decimals	2	5	1
Equivalent fractions	2	6	1
Multiplication: money	1 & 2	7	1
Capacity	3	8	1
Perimeter	3	9	1
2-D shape/symmetry	3	10	1
Fraction of amounts	2	11	1
Multiplication	2	12	1
Time	3	13	1
Temperature difference	2	14	1
3-D shape	3	15	1
Multiplication: decimals	2	16	1
Doubling decimals	2	17	1
Rounding after division: money	1 & 2	18	1
Difference: near multiple of 1000	3	19	1
Multiplying/dividing whole numbers by 100	3	20	1

Mental mathematics test score (out of 20)

Name

1.
2.
3.
4.
5.
6.
7.
8.
9.
10.
11.
12.
13.
14.
15.
16.
17.
18.
19.
20.
21.
22.
23.
24.
25.
26.
27.
28.
29.
30.

Number correct

Number incorrect or omitted

Percentage correct

Percentage incorrect or omitted

Record-keeping format 4 Test 3 grid for test analysis (Paper A)

Topic	Level	Question	Marks
Rounding and ordering decimals	2	1	1
Rounding decimals	2	2	1
Multiplying whole numbers and decimals by 100	2	3	1
Subtraction: money	2	4	1
Ordering fractions and decimals	2	5	1
Multiplication (inverse)	2	6	1
Ordering decimals	2	7	1
Time: 24-hour notation	3	8	1
Handling data: table and Venn diagram	4	9	2
Multiples	2	10	1
Multiplication and subtraction: money	1 & 2	11	2
Handling data: Carroll diagram	4	12ab	2
Division: capacity	1, 2 & 3	13	1
Time difference	1, 2 & 3	14	1
Perimeter	3	15	1
a) Division: money b) Addition: money	1 & 2	16ab	2
Area	3	17	1
a) Multiplication: capacity b) Multiplication: money	1, 2 & 3	18ab	2
Multiplication	1 & 2	19	2
3-D shape: nets	3	20	1
Handling data: table and Venn diagram	4	21	2
Handling data: line graph	4	22ab	2
Percentages of amounts: capacity	1, 2 & 3	23	2
Capacity: reading and interpreting scales	3	24	1
Angles in a straight line and in a triangle	3	25AB	2

Student rows (blank): 1.–30.

Paper A score (out of 35)

Record-keeping format 4 Test 3 grid for test analysis (Paper B)

Topic	NC Level	Question	Marks
Common multiples	2	1	1
Equivalent fractions	2	2	2
Ordering whole numbers	2	3	1
Addition: money	2	4	1
Percentages of numbers	2	5	1
Capacity: calculating with scales	1, 2 & 3	6	1
Multiplying decimals by 100	2	7	1
Addition: fractions	2	8	1
Addition (inverse)	2	9	1
Factors	2	10	2
Understanding word problems	1 & 2	11	1
Multiplication (inverse)	2	12	1
Algebra	1 & 2	13ab	2
Calculating: time	1, 2 & 3	14	2
Length: subtraction and division (inverse)	1, 2 & 3	15ab	2
Area	3	16	1
Time: calendar	3	17	2
Angles in a triangle	3	18	2
Properties of 3-D solids	3	19	2
Handling data: bar chart	4	20ab	2
a) Percentages of amounts; b) Division capacity	1, 2 & 3	21ab	2
Co-ordinates/2-D shapes/ Symmetry	1, 2 & 3	22ab	2
Handling data: Venn diagram	4	23	1
Rotation	3	24	1

Additional columns: Paper B score (out of 35) · Total score (out of 90) · National Curriculum Level

Rows numbered 1.–30. (blank)

Record-keeping format 5 Attainment Target 1 – Using and applying mathematics

Level 2

Problem solving
- Select and use material in some classroom activities
- Select and use mathematics for some classroom activities
- Begin to develop own strategies for solving a problem
- Begin to understand ways of working through a problem

Communicating
- Discuss work using mathematical language
- Respond to and ask mathematical questions
- Begin to represent work using symbols and simple diagrams
- Explain why an answer is correct

Reasoning
- Ask questions such as: 'What would happen if…?' 'Why?'
- Begin to develop simple strategies

Level 3

Problem solving
- Develop different mathematical approaches to a problem
- Look for ways to overcome difficulties
- Begin to make decisions and realise that results may vary according to the 'rule' used
- Begin to organise work
- Check results

Communicating
- Discuss mathematical work
- Begin to explain thinking
- Use and interpret mathematical symbols and diagrams

Reasoning
- Understand a general statement
- Investigate general statements and predictions by finding and trying out examples

Level 4

Problem solving
- Develop own strategies for solving problems
- Use own strategies for working within mathematics
- Use own strategies for applying mathematics to practical contexts

Communicating
- Present information and results in a clear and organised way

Reasoning
- Search for solutions by trying out own ideas

Level 5

Problem solving
- Identify and obtain necessary information
- Check results, considering whether these are sensible

Communicating
- Show understanding of situations by describing them mathematically using symbols, words and diagrams

Reasoning
- Draw simple conclusions
- Give an explanation for their reasoning

General comments

Year 5 National Expectations
Start-of-year: Level 3b
End-of-year: Level 3a (4c)

177

Record-keeping format 6 Attainment Target 2 – Number and algebra

Level 2

Numbers and the number system	Calculations	Solving numerical problems
• Count sets of objects reliably • Understand place value (HTU) • Order numbers up to 100 • Recognise sequences of numbers • Recognise odd and even numbers	• Recall addition and subtraction number facts to 10 • Understand that subtraction is the inverse of addition	• Use appropriate operation • Use mental strategies to solve problems involving money and measures

Level 3

Numbers and the number system	Calculations	Solving numerical problems
• Understand place value (ThHTU) • Begin to use decimal notation • Recognise negative numbers • Use simple fractions that are several parts of a whole • Recognise when two fractions are equivalent	• Make approximations • Recall addition and subtraction number facts to 20 • Add and subtract two two-digit numbers mentally • Add and subtract three two-digit numbers using written methods • Recall 2, 3, 4, 5, 10 multiplication tables • Recall division facts corresponding to the 2, 3, 4, 5, 10 multiplication tables	• Solve word problems involving larger numbers • Solve word problems involving multiplication • Solve word problems involving division, including those with a remainder

Level 4

Numbers and the number system	Calculations	Solving numerical problems
• Multiply and divide whole numbers by 10 or 100 • Add and subtract decimals to two places • Order decimals to three places • Recognise approximate proportions of a whole • Use simple fractions and percentages to describe proportions of a whole • Recognise and describe number patterns • Recognise and describe a multiple, factor and square	• Recall multiplication facts up to 10 × 10 • Recall division facts corresponding to the multiplication facts up to 10 × 10 • Use efficient written methods for addition and subtraction • Use efficient written methods for short multiplication and division	• Use a range of mental calculation strategies for the four operations • Check the reasonableness of an answer • Begin to use simple formulae expressed in words • Use and interpret co-ordinates in the first quadrant

Level 5

Numbers and the number system	Calculations	Solving numerical problems
• Multiply and divide whole numbers and decimals by 10, 100 and 1000 • Order negative numbers • Add and subtract negative numbers • Reduce a fraction to its simplest form • Solve simple problems involving ratio and proportion • Calculate fractional or percentage parts of quantities and measurements	• Use brackets appropriately • Use efficient written methods for addition and subtraction up to 10 000 • Use efficient written methods for long multiplication and division • Use all four operations with decimals to two places	• Check solutions by applying inverse operations • Check solutions by estimating using approximations • Construct and express in symbolic form simple formulae involving one or two operations • Use and interpret co-ordinates in all four quadrants

General comments

Year 5 National Expectations

Start-of-year: Level 3b

End-of-year: Level 3a (4c)

Record-keeping format 7 Attainment Target 3 – Shape, space and measures

Level 2

Understanding properties of shapes
- Use mathematical names for common 2-D and 3-D shapes
- Describe the properties of common 2-D and 3-D shapes, including number of sides and corners

Understanding properties of position and movement
- Distinguish between straight and turning movements
- Understand angle as a measure of turn
- Recognise right angles in turns

Understanding measures
- Begin to use everyday non-standard units to measure length and mass
- Begin to use everyday standard units to measure length and mass

Level 3

Understanding properties of shapes
- Classify 3-D and 2-D shapes in various ways

Understanding properties of position and movement
- Use mathematical properties such as reflective symmetry to describe 2-D shapes

Understanding measures
- Use non-standard units of length, capacity and mass in a range of contexts
- Use standard units of length, capacity, mass and time in a range of contexts

Level 4

Understanding properties of shapes
- Make 3-D mathematical models by linking given faces or edges
- Draw common 2-D shapes in different orientations on grids

Understanding properties of position and movement
- Reflect simple shapes in a mirror line

Understanding measures
- Choose and use appropriate units
- Choose and use appropriate instruments
- Interpret, with appropriate accuracy, numbers on a range of measuring instruments
- Find perimeters of simple shapes
- Find areas by counting squares

Level 5

Understanding properties of shapes
- Measure and draw angles to the nearest degree when constructing models and drawing shapes

Understanding properties of position and movement
- Use language associated with angle
- Know the angle sum of a triangle
- Know the sum of angles at a point
- Identify all the symmetries of 2-D shapes

Understanding measures
- Know the rough metric equivalences of imperial units still in daily use
- Convert one metric unit to another
- Make sensible estimates of a range of measures in relation to everyday situations
- Understand and use the formula for the area of a rectangle

General comments

Record-keeping format 8 Attainment Target 4 – Handling data

Level 2

Processing, representing and interpreting data
- Sort objects using more than one criterion
- Classify objects using more than one criterion
- Record results in simple lists, tables and block graphs
- Communicate findings

Level 3

Processing, representing and interpreting data
- Extract and interpret information presented in simple tables and lists
- Construct and interpret bar charts and pictograms where the symbol represents a group of units

Level 4

Processing, representing and interpreting data
- Collect discrete data and record them using a frequency table
- Understand and use the mode and range to describe sets of data
- Group data, where appropriate, in equal class intervals
- Represent collected data in frequency diagrams and interpret such diagrams
- Construct and interpret simple line graphs

Level 5

Processing, representing and interpreting data
- Understand and use the mean of discrete data
- Compare two simple distributions using the range and one of the mode, median or mean
- Interpret graphs and diagrams, including pie charts, and draw conclusions
- Understand and use the probability scale from 0 to 1
- Find and justify probabilities
- Select and use methods based on equally likely outcomes and experimental evidence
- Understand that different outcomes may result from repeating an experiment

General comments

Year 5 National Expectations

Start-of-year: Level 3b

End-of-year: Level 3a (4c)

Record-keeping format 9 Class record of the end-of-year expectations

Class: _____

Date: _____

Names

Year 5
End-of-year expectations

Counting and understanding number (AT2)
Explain what each digit represents in whole numbers and decimals with up to two places, and partition, round and order these numbers (Level 4)

Knowing and using number facts (AT2)
Use knowledge of place value and addition and subtraction of two-digit numbers to derive sums and differences and doubles and halves of decimals, e.g. 6·5 ± 2·7, halve 5·6, double 0·34 (Level 4)

Calculating (AT2)
Use efficient written methods to add and subtract whole numbers and decimals with up to two places (Level 4)

Understanding shape (AT3)
Read and plot co-ordinates in the first quadrant; recognise parallel and perpendicular lines in grids and shapes; use a set-square and ruler to draw shapes with perpendicular or parallel sides (Level 4)

Measuring (AT3)
Draw and measure lines to the nearest millimetre; measure and calculate the perimeter of regular and irregular polygons; use the formula for the area of a rectangle to calculate the rectangle's area (Level 4)

Handling data (AT4)
Construct frequency tables, pictograms and bar and line graphs to represent the frequencies of events and changes over time (Level 4)

NOTES: **Using and applying mathematics (AT1)** is incorporated throughout
End-of-year National Expectations: Level 3a (4c)

Record-keeping format 10 Individual child's record of the end-of-year expectations

Name: _____

Foundation Stage	Year 1	Year 2	Year 3
Using and applying mathematics (AT1)			
Use developing mathematical ideas and methods to solve practical problems (Level 1)			
Talk about, recognise and recreate simple patterns (Level 1)			
Counting and understanding number (AT2)			
Say and use the number names in order in familiar contexts (Level 1)	Read and write numerals from 0 to 20, then beyond; use knowledge of place value to position these numbers on a number track and number line (Level 2)	Count up to 100 objects by grouping them and counting in tens, fives or twos; explain what each digit in a two-digit number represents, including numbers where 0 is a place holder; partition two-digit numbers in different ways, including into multiples of 10 and 1 (Level 2)	Partition three-digit numbers into multiples of 100, 10 and 1 in different ways (Level 3)
Count reliably up to 10 everyday objects (Level 1)			
Use language such as 'more' or 'less' to compare two numbers (Level 1)			
Recognise numerals 1 to 9 (Level 1)			
Knowing and using number facts (AT2)			
Find one more or one less than a number from 1 to 10 (Level 1)	Derive and recall all pairs of numbers with a total of 10 and addition facts for totals to at least 5; work out the corresponding subtraction facts (Level 2)	Derive and recall all addition and subtraction facts for each number to at least 10, all pairs with totals to 20 and all pairs of multiples of 10 with totals up to 100 (Level 2)	Derive and recall all addition and subtraction facts for each number to 20, sums and differences of multiples of 10 and number pairs that total 100 (Level 3)
Calculating (AT2)			
Begin to relate addition to combining two groups of objects and subtraction to 'taking away' (Level 1)	Use the vocabulary related to addition and subtraction and symbols to describe and record addition and subtraction number sentences (Level 2)	Add or subtract mentally a one-digit number or a multiple of 10 to or from any two-digit number; use practical and informal written methods to add and subtract two-digit numbers (Level 2)	Add or subtract mentally combinations of one-digit and two-digit numbers (Level 3)
In practical activities and discussion begin to use the vocabulary involved in adding and subtracting (Level 1)		Use the symbols $+$, $-$, $\times$, $\div$ and $=$ to record and interpret number sentences involving all four operations; calculate the value of an unknown in a number sentence (Level 2)	
Understanding shape (AT3)			
Use language such as 'circle' or 'bigger' to describe the shape and size of solids and flat shapes (Level 1)	Visualise and name common 2-D shapes and 3-D solids and describe their features; use them to make patterns, pictures and models (Level 2)	Visualise common 2-D shapes and 3-D solids; identify shapes from pictures of them in different positions and orientations; sort, make and describe shapes, referring to their properties (Level 2)	Draw and complete shapes with reflective symmetry and draw the reflection of a shape in a mirror line along one side (Level 3)
Use everyday words to describe position (Level 1)			
Measuring (AT3)			
Estimate, measure, weigh and compare objects, choosing and using suitable uniform non-standard or standard units and measuring instruments, e.g. a lever balance, metre stick or measuring jug (Level 2)		Use units of time (seconds, minutes, hours, days) and know the relationships between them; read the time to the quarter hour; identify time intervals, including those that cross the hour (Level 2)	Read, to the nearest division and half-division, scales that are numbered or partially numbered; use the information to measure and draw to a suitable degree of accuracy (Level 3)
Use language such as 'greater', 'smaller', 'heavier' or 'lighter' to compare quantities (Level 1)			
Handling data (AT4)			
Answer a question by recording information in lists and tables; present outcomes using practical resources, pictures, block graphs or pictograms (Level 2)		Use lists, tables and diagrams to sort objects; explain choices using appropriate language, including 'not' (Level 2)	Use Venn diagrams or Carroll diagrams to sort data and objects using more than one criterion (Level 3)

NOTE: **Using and applying mathematics (AT1)** is incorporated throughout

Record-keeping format 10   Individual child's record of the end-of-year expectations   Name: _____

Year 4	Year 5	Year 6	Year 6 progression to Year 7
Counting and understanding number (AT2)			
Use diagrams to identify equivalent fractions; interpret mixed numbers and position them on a number line (Level 3)	Explain what each digit represents in whole numbers and decimals with up to two places, and partition, round and order these numbers (Level 4)	Express one quantity as a percentage of another; find equivalent percentages, decimals and fractions (Level 4)	Use ratio notation, reduce a ratio to its simplest form and divide a quantity into two parts in a given ratio; solve simple problems involving ratio and direct proportion (Level 5)
Knowing and using number facts (AT2)			
Derive and recall multiplication facts up to 10 × 10, the corresponding division facts and multiples of numbers to 10 up to the tenth multiple (Level 4)	Use knowledge of place value and addition and subtraction of two-digit numbers to derive sums and differences and doubles and halves of decimals (Level 4)	Use knowledge of place value and multiplication facts to 10 × 10 to derive related multiplication and division facts involving decimals (Level 4)	Make and justify estimates and approximations to calculations (Level 5)
Calculating (AT2)			
Add or subtract mentally pairs of two-digit whole numbers (Level 3) Develop and use written methods to record, support and explain multiplication and division of two-digit numbers by a one-digit number, including division with remainders (Level 4)	Use efficient written methods to add and subtract whole numbers and decimals with up to two places (Level 4)	Use efficient written methods to add and subtract integers and decimals, to multiply and divide integers and decimals by a one-digit integer, and to multiply two-digit and three-digit integers by a two-digit integer (Level 5)	Use bracket keys and the memory of a calculator to carry out calculations with more than one step; use the square-root key (Level 5)
Understanding shape (AT3)			
Know that angles are measured in degrees and that one whole turn is 360°; compare and order angles less than 180° (Level 3)	Read and plot co-ordinates in the first quadrant; recognise parallel and perpendicular lines in grids and shapes; use a set-square and ruler to draw shapes with perpendicular or parallel sides (Level 4)	Visualise and draw on grids of different types where a shape will be after reflection, after translations, or after rotation through 90° or 180° about its centre or one of its vertices (Level 5)	Know the sum of angles on a straight line, in a triangle and at a point, and recognise vertically opposite angles (Level 5)
Measuring (AT3)			
Choose and use standard metric units and their abbreviations when estimating, measuring and recording length, weight and capacity; know the meaning of 'kilo', 'centi' and 'milli' and, where appropriate, use decimal notation to record measurements (Level 3) Draw and measure lines to the nearest millimetre; measure and calculate the perimeter of regular and irregular polygons; use the formula for the area of a rectangle to calculate the rectangle's area (Level 4)		Select and use standard metric units of measure and convert between units using decimals to two places (Level 4)	Solve problems by measuring, estimating and calculating; measure and calculate using imperial units still in everyday use; know their approximate metric values (Level 5)
Handling data (AT4)			
Answer a question by identifying what data to collect; organise, present, analyse and interpret the data in tables, diagrams, tally charts, pictograms and bar charts, using ICT where appropriate (Level 3)	Construct frequency tables, pictograms and bar and line graphs to represent the frequencies of events and changes over time (Level 4)	Solve problems by collecting, selecting, processing, presenting and interpreting data, using ICT where appropriate; draw conclusions and identify further questions to ask (Level 5)	Understand and use the probability scale from 0 to 1; find and justify probabilities based on equally likely outcomes in simple contexts (Level 5)

NOTE: **Using and applying mathematics (AT1)** is incorporated throughout

	Foundation Stage	Year 1	Year 2	Year 3	Year 4	Year 5	Year 6
End-of-year National Expectations	1b	1a (2c)	2b	2a (3c)	3b	3a (4c)	4b

Word problem cards

1. During the summer, Brian waters his plants every day. He uses one 2·5 litre watering can of water. How much water does he use in a 7 day week in litres and millilitres?

2. A theatre has 396 seats in the stalls, 356 seats in the dress circle and 228 seats in the balcony. How many seats is this altogether?

3. Coloured pencils come in packets of 10. There are 14 packets to a box. A school buys 6 boxes and shares them equally between 12 classes. How many packets does each class get?

4. Fina is training for the marathon. Each day she spends 3 hours training. She runs 10 times around a 600 m track. How many kilometres does she run each day?

5. Oranges cost £1.12 per kg. Yolanda buys 6 kg. What is the total cost of the 6 kg of oranges?

6. Jenny wins the 100 metres running race in 22 seconds. How long does she take to run 10 metres?

7. Katie gets the 06:45 bus to the station. She arrives at the station at 07:10. She then gets on a train that takes her 1 hr and 40 min. How long was Katie's journey?

8. Lane Primary School has bought 6 bottles of squash for their sports day. Each bottle holds 1·5 litres. If 800 ml have been used, in litres and millilitres, how much squash is left?

9. Charlie and Alex's aunt gives them £60 to share equally between the two of them. Charlie spends one half of his share, and Alex spends one quarter of his share. How much has Charlie left? How much has Alex left?

10. Brian buys 4·5 kg of potatoes, 3·2 kg of carrots, 2·8 kg of beans and 1·4 kg of onions. How many kilograms and grams of vegetables does he buy altogether?

11. During their Summer Sale, Better Electrics are offering 25% discount on all TVs. How much does a TV normally priced at £120 cost during the sale?

12. Steve is buying a motorbike for £8558. He pays a deposit of £855.80. How much does he have left to pay when he collects the motorbike?

13. I am thinking of a number. If you divide it by 6 and then multiply by 9, you get 63. What number am I thinking of?

14. A farmer owns a plot of land 15·5 m by 12 m. What is the size of the fence around the perimeter of the land?

15. Mr. Keft grows 40 acres of potatoes. Each acre produces 650 kg of potatoes. If he sells potatoes in 2·5 kg bags, how many bags of potatoes does he have to sell?

16. Mr. and Mrs. Yates have just bought two rectangular plots of land. One is 14 m by 18 m, the other 16 m by 9 m. What is the total area of the two plots of land?

17. I am thinking of a number. If you add 8 to this number and then multiply by 100 and subtract 680, you get 320. What number am I thinking of?

18. Buffy and 5 friends go to the cinema. Buffy pays for them all with £50 and receives £11 change. How much was each ticket?

19. Peter takes the ferry to the island of St. Honorat. The ferry leaves at 11:15 and arrives at the island 25 minutes later. If he leaves the island on the 5:30 ferry, how long does he spend on the island?

20. The Nowra Tennis Club has 3 adult members for every 5 junior members. If there are 60 adult members, how many junior members are there?

© Collins New Primary Maths

Puzzles 1

The sum of three even numbers is always even.

The sum of three odd numbers is always even.

The difference between one odd and one even number is always odd.

The product of two even numbers is always even.

The product of two odd numbers is always odd.

The difference between two odd or two even numbers is always odd.

The product of one odd and one even number is always odd.

- Investigate which of Jennifer's statements are true.

- Be sure to give examples to prove your conclusions.

The number 162 is divisible by 3

The number 155 is divisible by 5

The number 146 is divisible by 4

The number 158 is divisible by 6

- Do you agree with Martin's statements? Why?

- Be sure to give examples to prove your conclusions.

- Make a general statement about numbers that are divisible by 3, 4, 5 and 6.

Collins New Primary Maths

Puzzles 2

- This clock is set correctly at 4 p.m.

- It loses four minutes every hour.

- What time will this clock read when the correct time is 11 a.m. the next day?

- Imagine a regular pentagon with all its diagonals drawn.

- How many triangles are there?

C Collin
New
Primar
Maths

Two-digit number cards

14	17	23	26
32	38	41	45
50	59	62	64
67	71	75	76
80	89	93	96

Collins New Primary Maths

Three-digit number cards

109	187	243	285
326	371	458	476
515	560	604	639
692	726	748	797
832	863	917	958

Four-digit number cards

1096	1554	2313	2862
3025	3648	4677	4981
5135	5306	6189	6491
6740	7265	7457	8536
8824	8973	9276	9782

Collins
New
Primary
Maths

Decimal cards – tenths

0·2	0·4	0·7	1·3
1·5	1·7	2·4	2·8
3·1	3·9	4·3	4·6
5·2	5·5	6·5	6·9
7·4	7·8	8·6	9·1

Decimal cards – hundredths

0·12	0·25	0·62	1·02
1·23	1·63	2·17	2·75
3·28	3·41	4·64	4·79
5·35	5·96	6·45	6·84
7·51	7·57	8·78	9·36

Collins New Primary Maths

Equivalent fraction cards

$\dfrac{1}{2}$	$\dfrac{2}{4}$	$\dfrac{3}{6}$	$\dfrac{4}{8}$
$\dfrac{1}{3}$	$\dfrac{2}{6}$	$\dfrac{3}{9}$	$\dfrac{4}{12}$
$\dfrac{1}{4}$	$\dfrac{2}{8}$	$\dfrac{3}{12}$	$\dfrac{4}{16}$
$\dfrac{1}{5}$	$\dfrac{2}{10}$	$\dfrac{3}{15}$	$\dfrac{4}{20}$
$\dfrac{1}{6}$	$\dfrac{2}{12}$	$\dfrac{3}{18}$	$\dfrac{4}{24}$
$\dfrac{3}{4}$	$\dfrac{6}{8}$	$\dfrac{2}{3}$	$\dfrac{4}{6}$

Collins New Primary Maths

Relating fractions to their decimal representations 1

1   2   3   4   5   8   10   12   100

0·04   0·1   0·25   0·02

0·2   0·05   0·8   0·75

0·03   0·3   0·01   0·12

0·4   0·5   0·08   0·6

Relating fractions to their decimal representations 2

Names _____ Date _____

1 2 3 4 5 8 10 12 100

| 0·12 | 1·6 | 0·03 | 0·8 | 0·3 |

| 2.5 | 0·1 | 0·6 | 0·5 | 0·75 | 1·5 |

| 0·2 | 0·08 | 0·04 | 1·25 | 0·25 |

1 2 3 4 5 6 7 8 9 10

| 2 | 0·8 | 0·3 | 0·1 | 3 |

| 0·5 | 0·25 | 0·4 | 7 | 0·6 |

| 0·9 | 0·2 | 0·7 | 0·75 | 4 |

Collins
New
Prima
Maths

Percentages 1

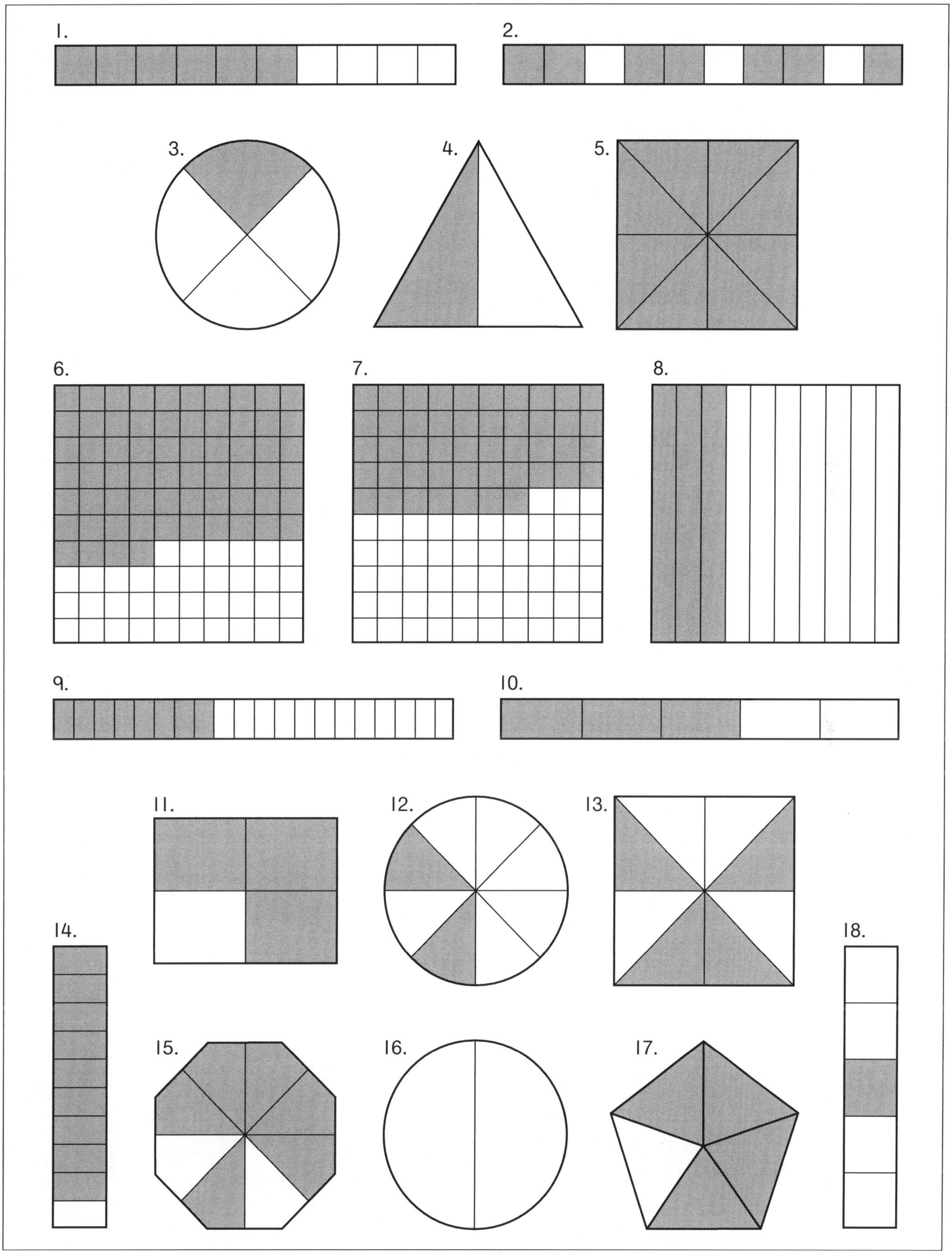

1.

2.

3.

4.

5.

6.

7.

8.

9.

10.

11.

12.

13.

14.

15.

16.

17.

18.

Collins
New
Primary
Maths

Proportion 1

1.
2.
3.
4.
5.
6.
7.
8.
9.
10.
11.
12.
13.
14.
15.
16.

Collins
New
Primary
Maths

Proportion 2

Chicken and Broccoli Stir-fry

(Serves 2 people)

200 g noodles
1 tablespoon oil
400 g chicken
$\frac{1}{2}$ large onion (100 g)
1 clove garlic
40 ml oyster sauce

Macaroni and Cheese

(Serves 4 people)

400 g macaroni
100 g butter
100 g plain flour
1 pint milk
260 g grated cheese
2 tablespoons oil

Sweet Peaches

(Serves 6 people)

6 peaches
2 tablespoons of caster sugar
3 drops of vanilla essence
300 g raspberries
60 g icing sugar
2 tablespoons of flaked almonds, toasted

Cheese Muffins

(Makes 10 muffins)

250 g plain flour
20 ml baking powder
50 ml sugar
100 g grated cheese
1 large egg
250 ml milk
100 ml corn oil

Collins New Primary Maths

Division facts

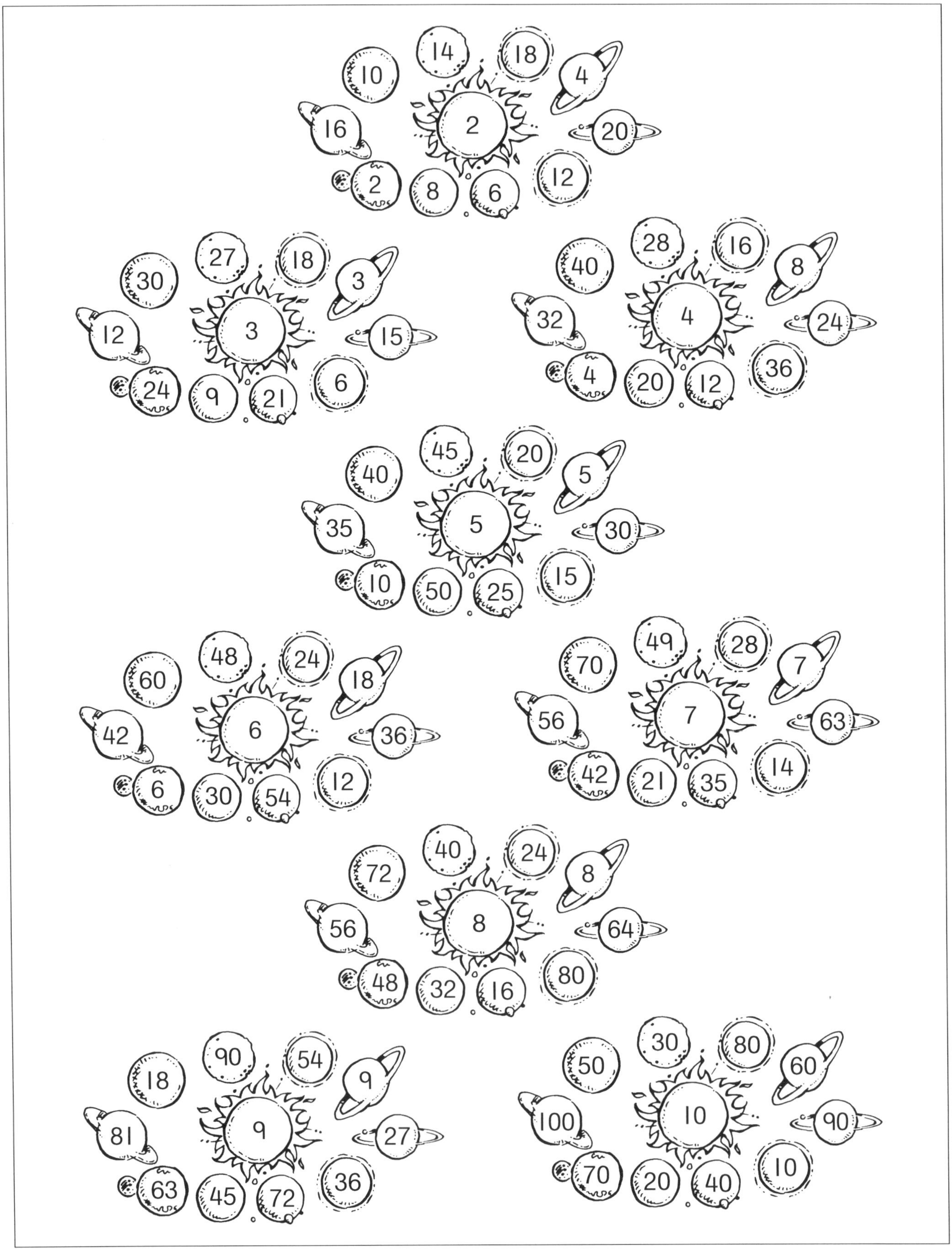
2
10
14
18
4
16
20
2
8
6
12
3
30
27
18
3
12
15
24
9
21
6
4
40
28
16
8
32
24
4
20
12
36
5
40
45
20
5
35
30
10
50
25
15
6
60
48
24
18
42
36
6
30
54
12
7
70
49
28
7
56
63
42
21
35
14
8
72
40
24
8
56
64
48
32
16
80
9
18
90
54
9
81
27
63
45
72
36
10
50
30
80
60
100
90
70
20
40
10

Estimate, calculate and check

Estimation	Calculation	Check

Collins New Primary Maths

Calculating a difference mentally

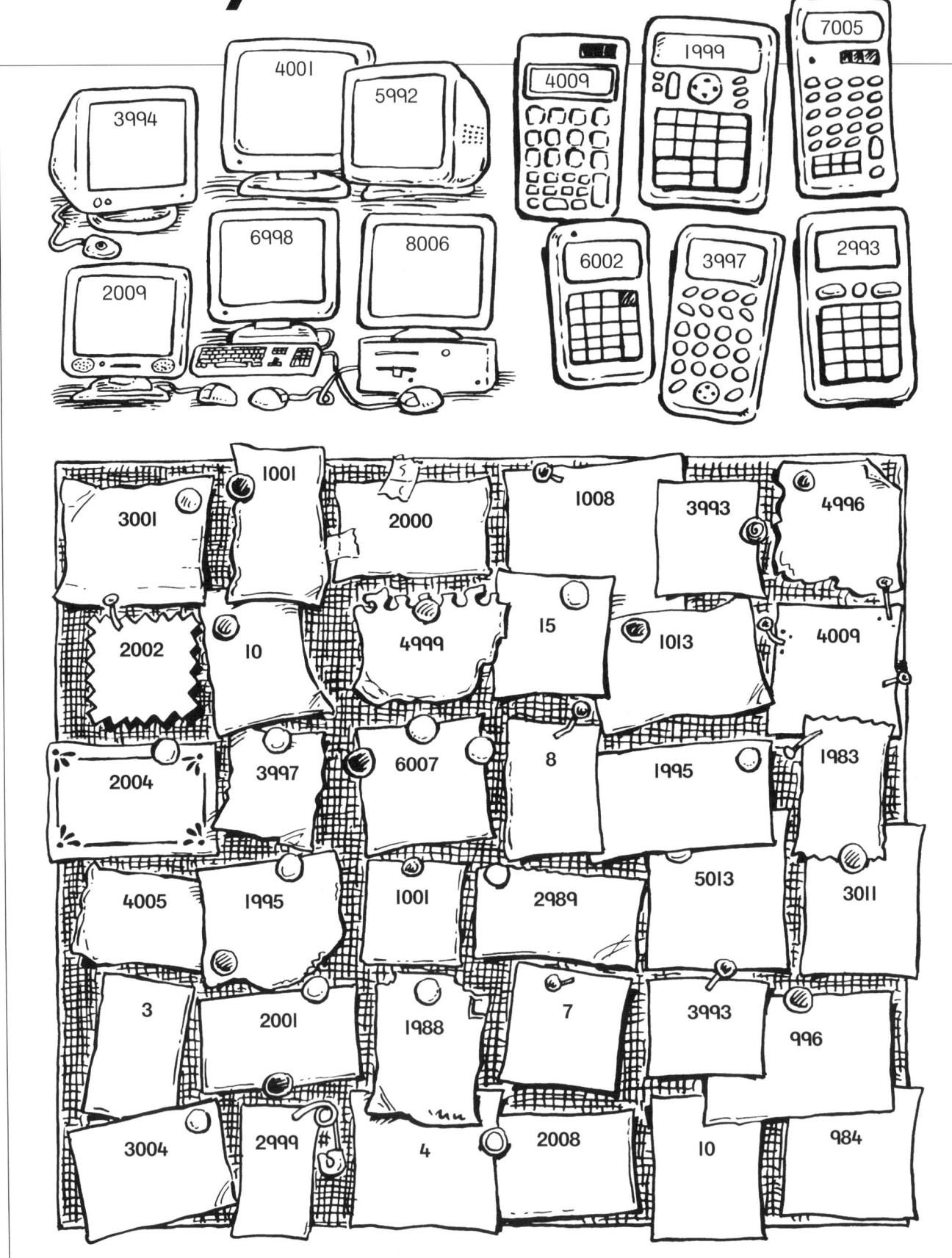

Finding fractions
of numbers

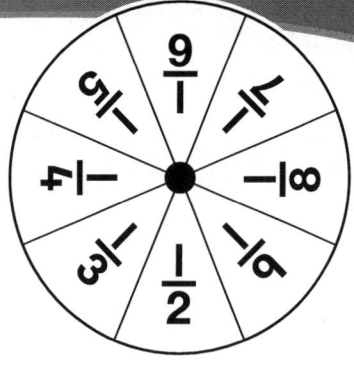

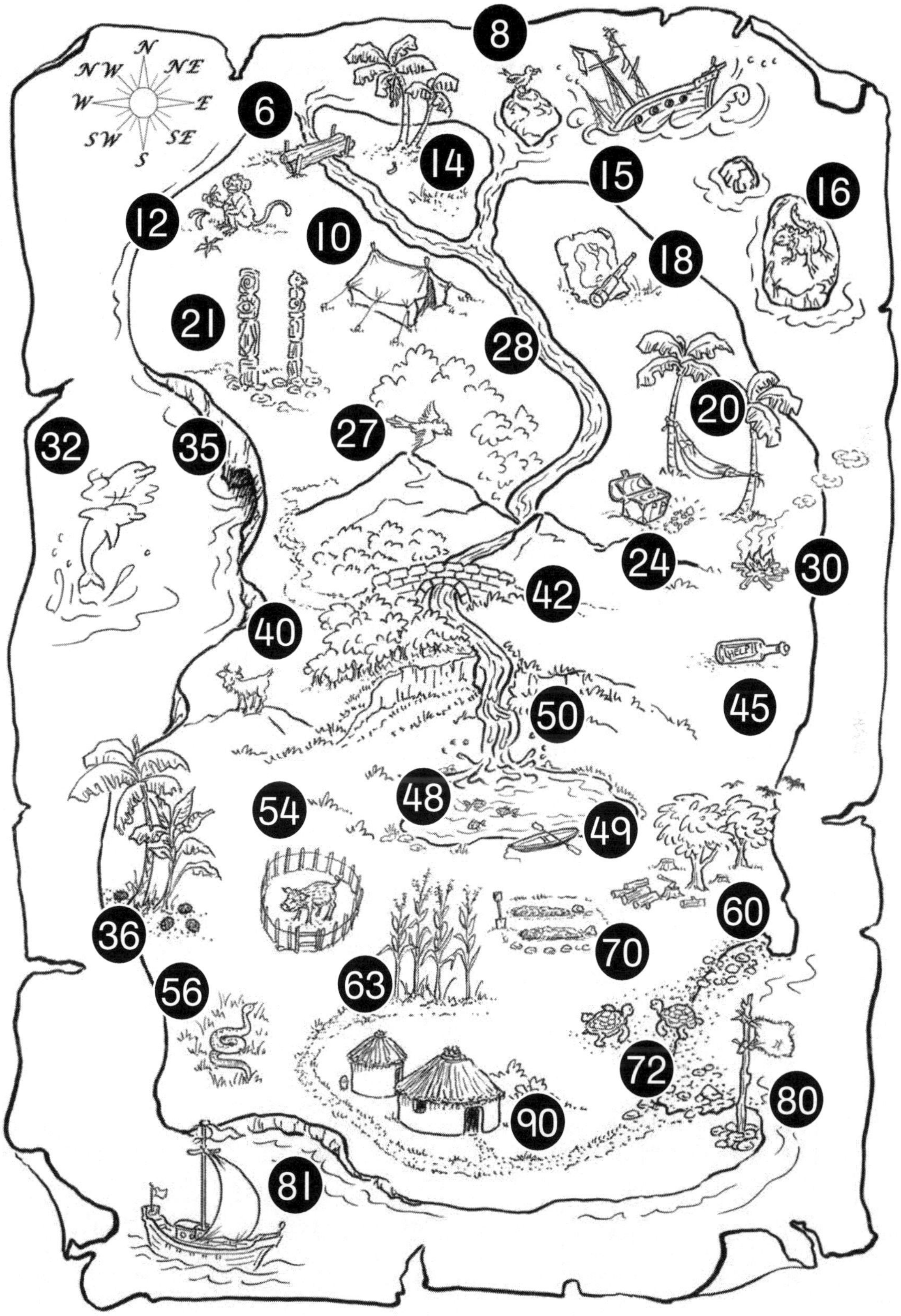

Spinner fractions: $\frac{1}{9}$, $\frac{1}{7}$, $\frac{1}{5}$, $\frac{1}{8}$, $\frac{1}{4}$, $\frac{1}{6}$, $\frac{1}{3}$, $\frac{1}{2}$

8

6   14   15   16

12   10   18

21   28   20

32   35   27   24   30

40   42   45

50

54   48   49

60

36   70

56   63   72   80

90

81

C. Collins New Primary Maths

Percentages 2

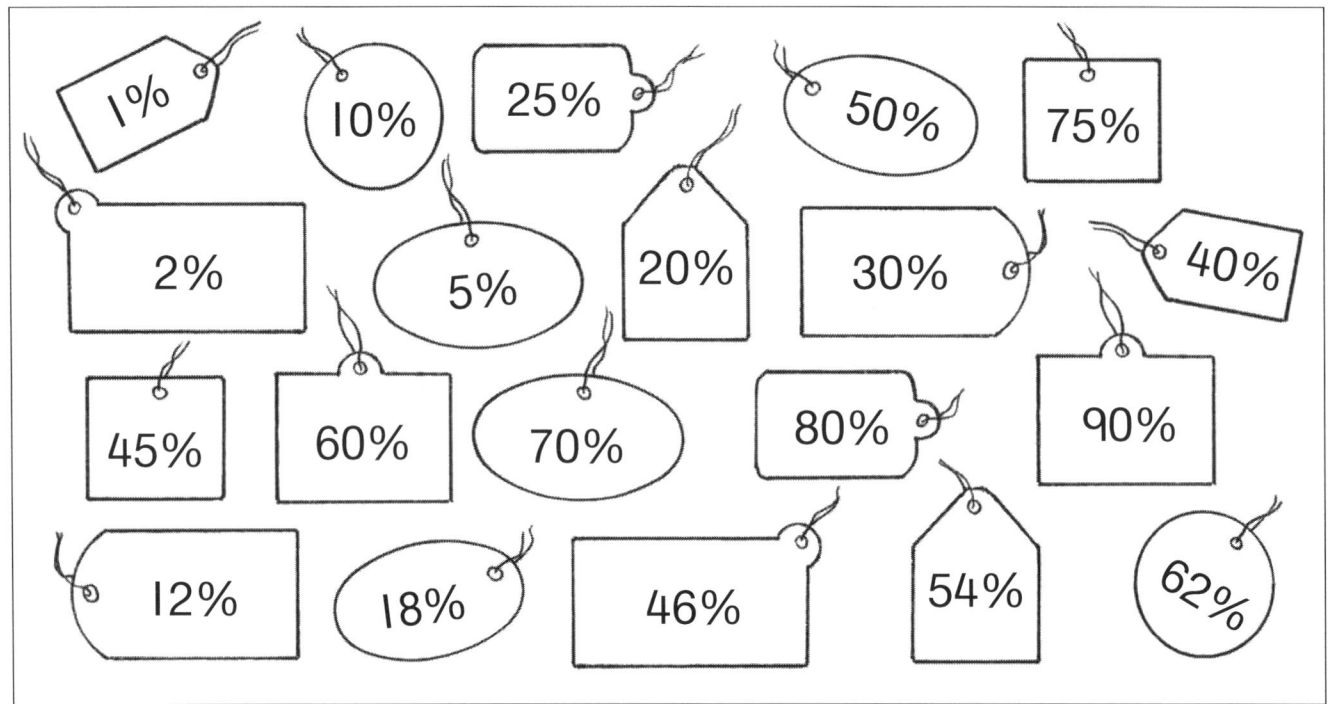

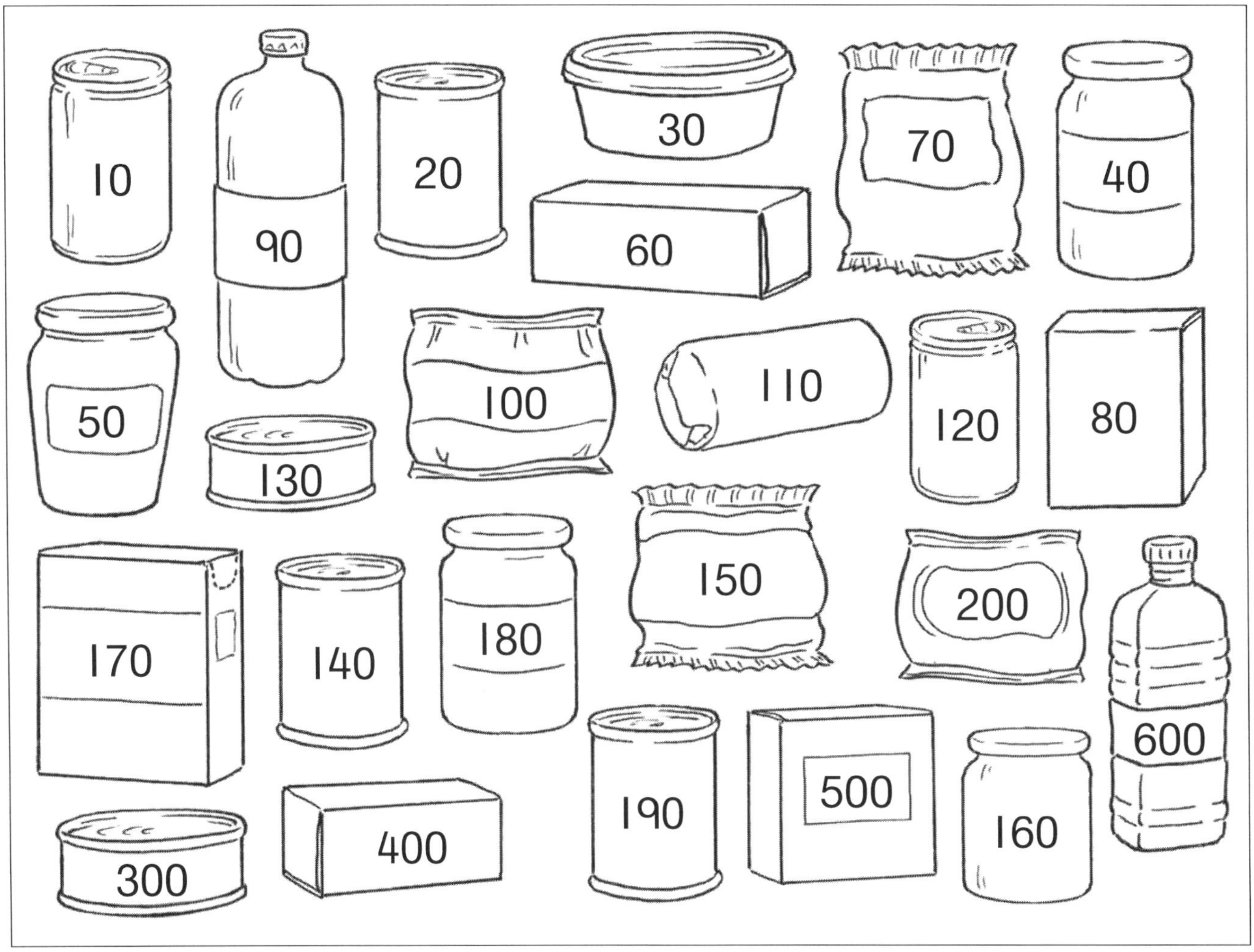

2-D shapes and 3-D solids

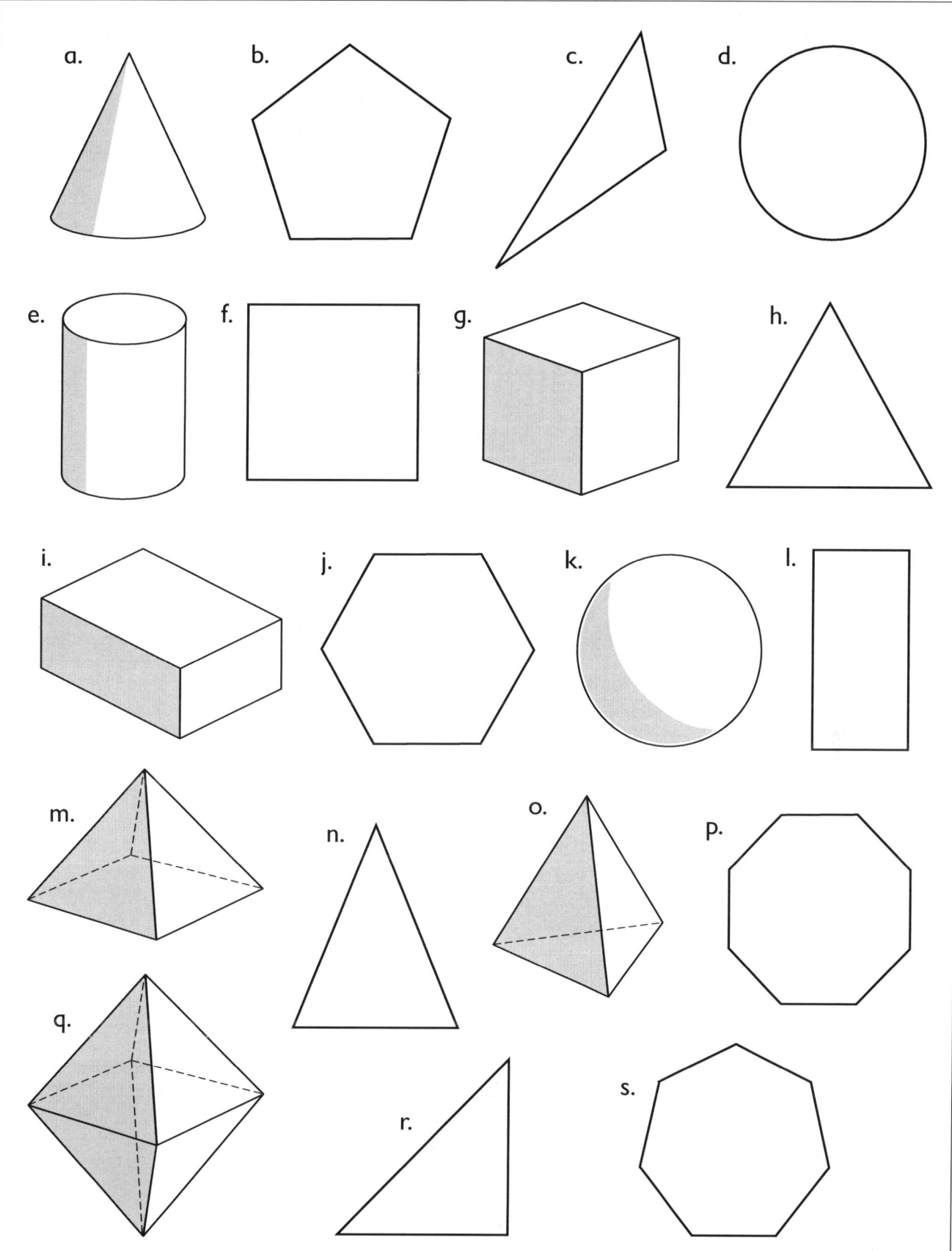

a.

b.

c.

d.

e.

f.

g.

h.

i.

j.

k.

l.

m.

n.

o.

p.

q.

r.

s.

Collins New Primary Maths

Nets

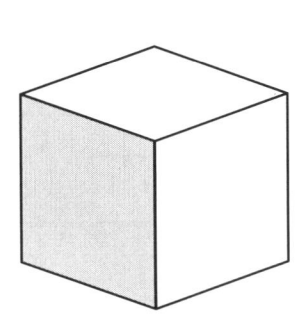

1. 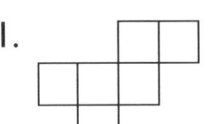 2. 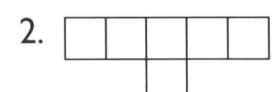   3.  4.

5. 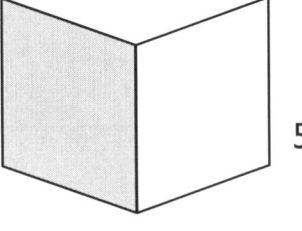 6. 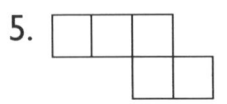 7. 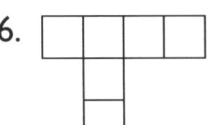 8.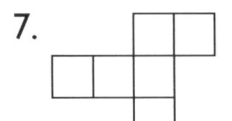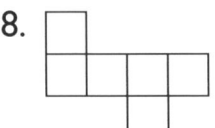

9. 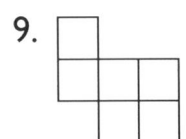 10. 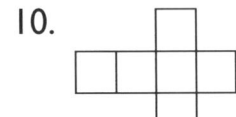 11. 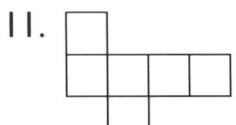 12.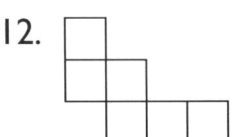

13. 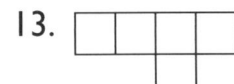 14. 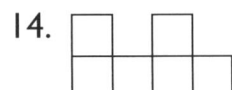 15. 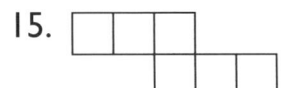 16.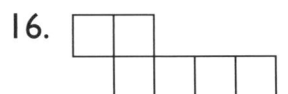

17. 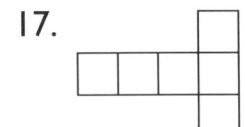 18. 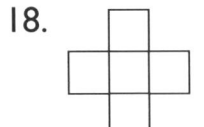 19. 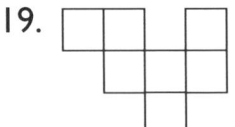 20.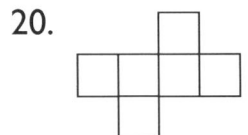

21. 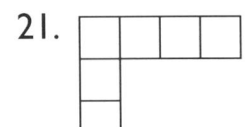 22. 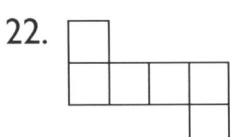 23. 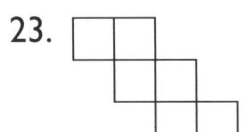 24.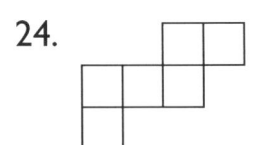

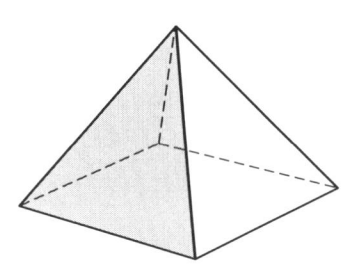

1. 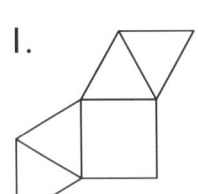 2. 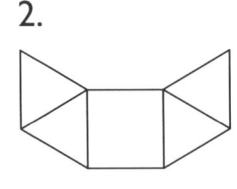 3. 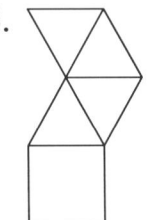 4.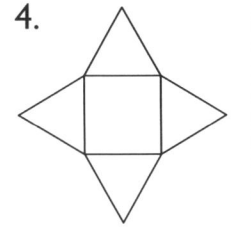

5. 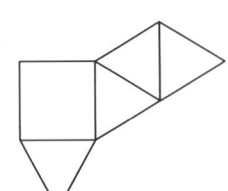 6. 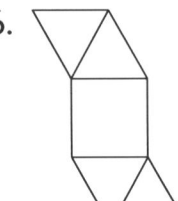 7. 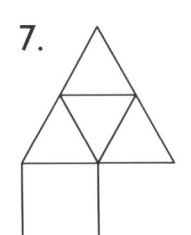 8.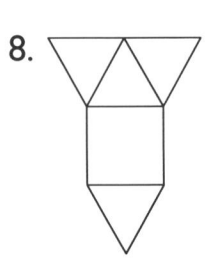

Co-ordinates and lines

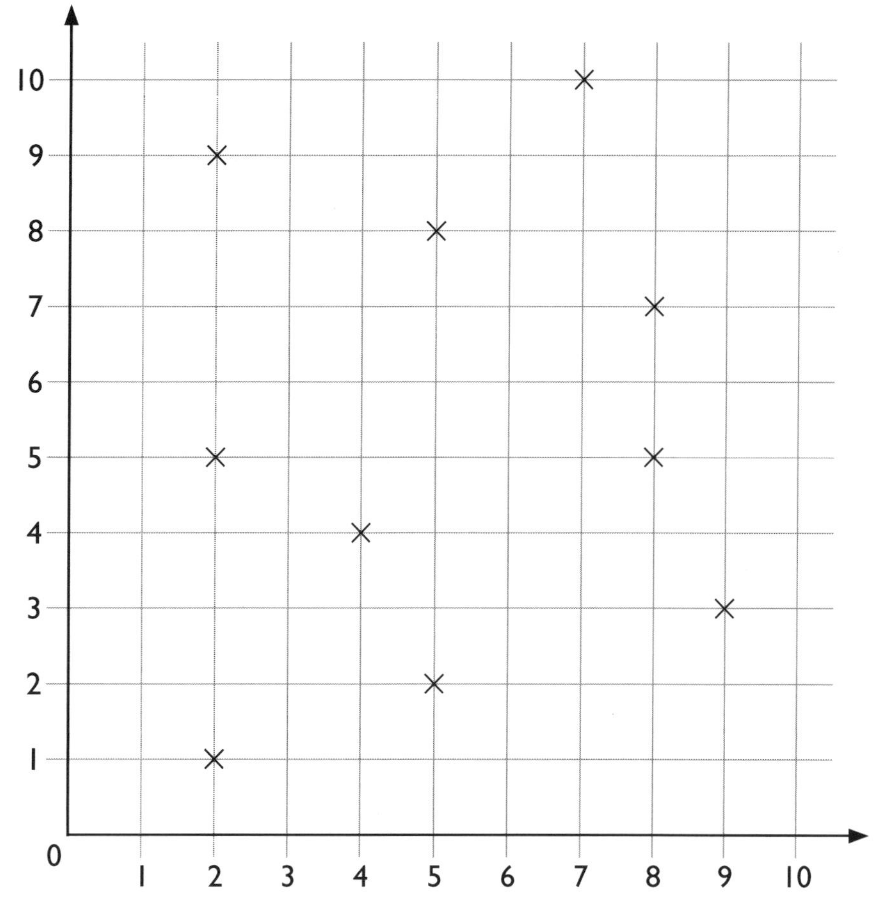

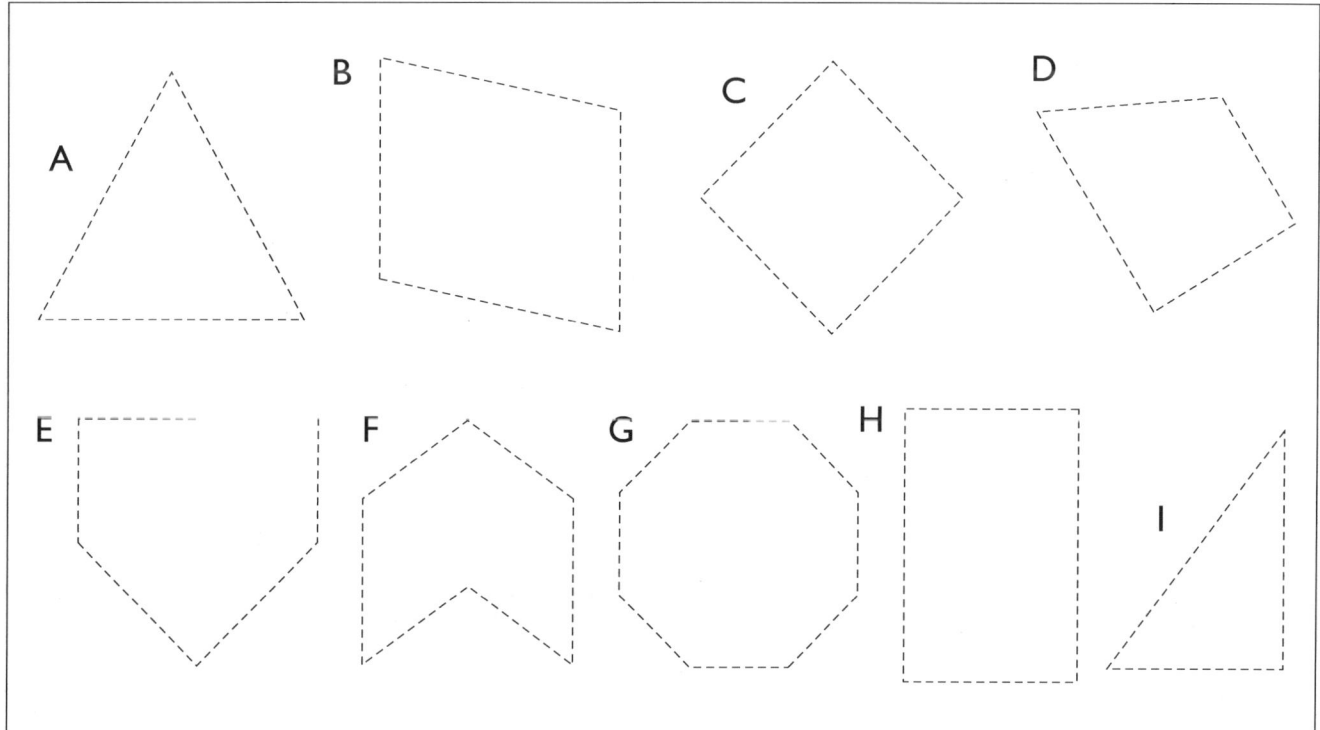

Symmetry, reflections and translations

Grid 1

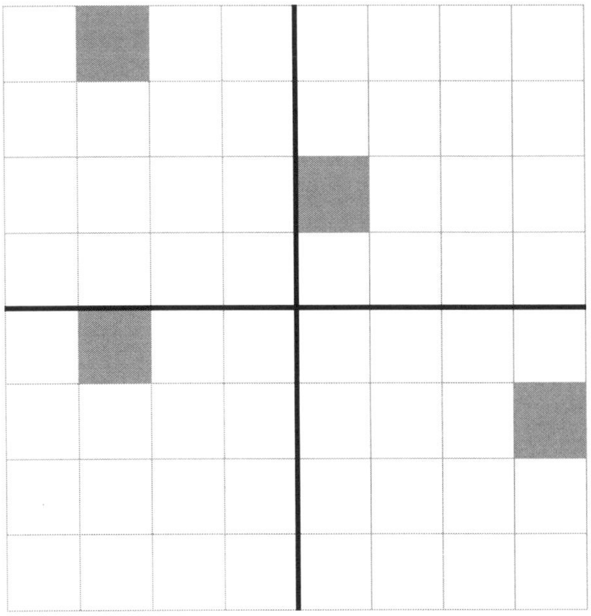

Grid 2

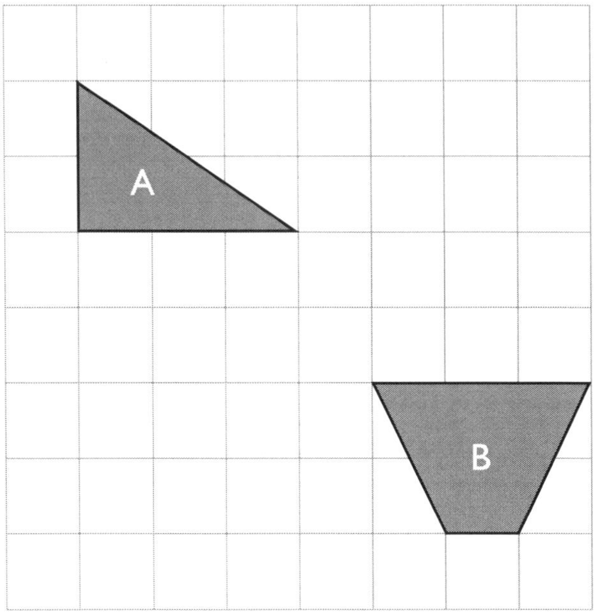

A

B

Grid 3

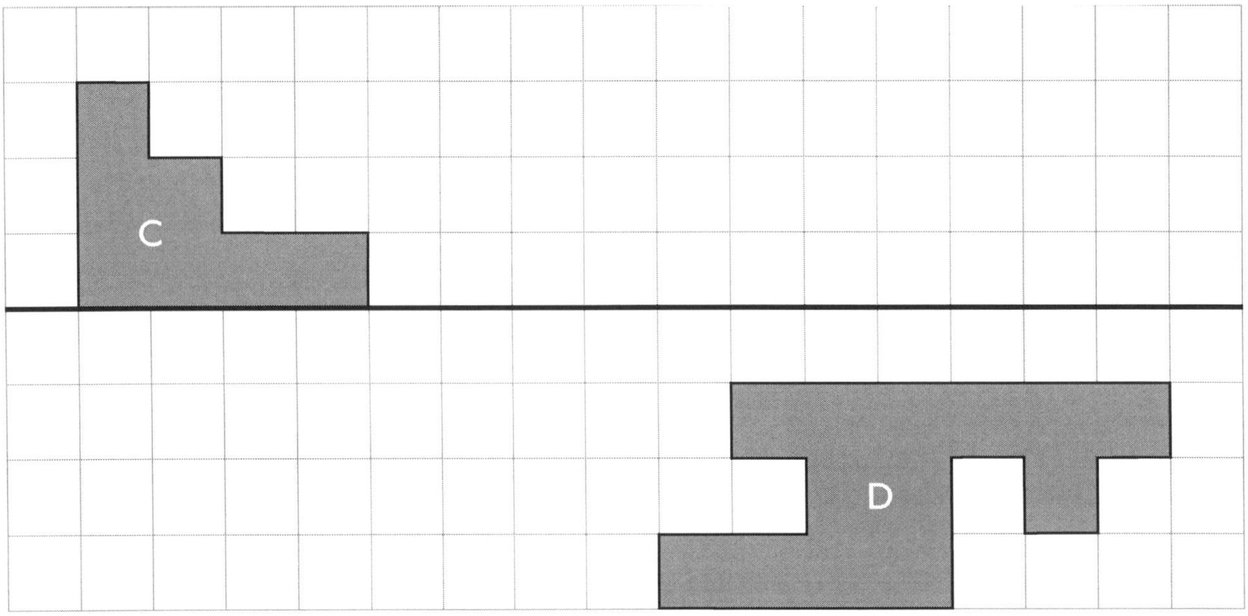

C

D

Collins
New
Primar
Maths

Measuring angle cards

1.

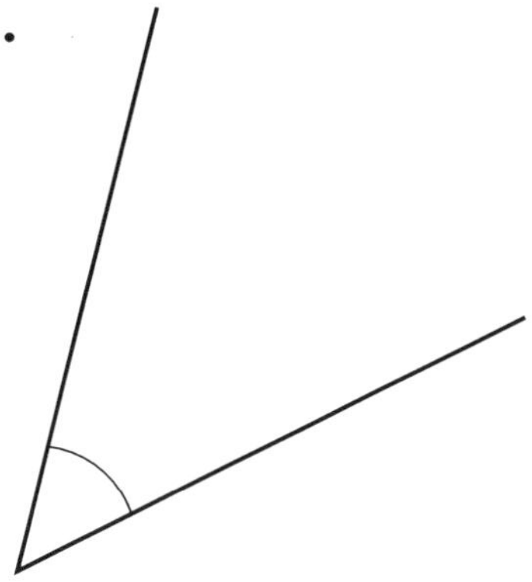

2.

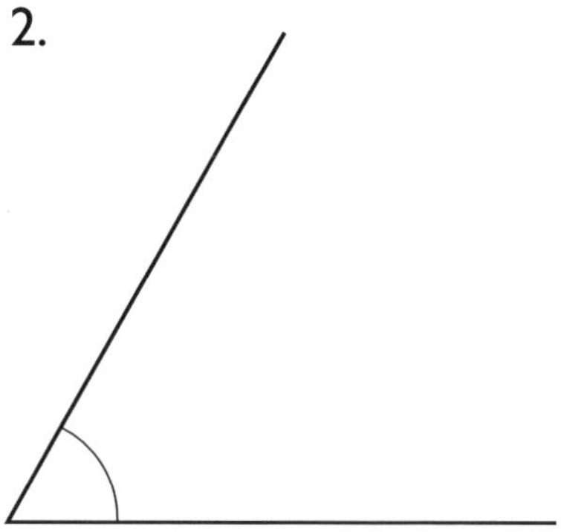

3.

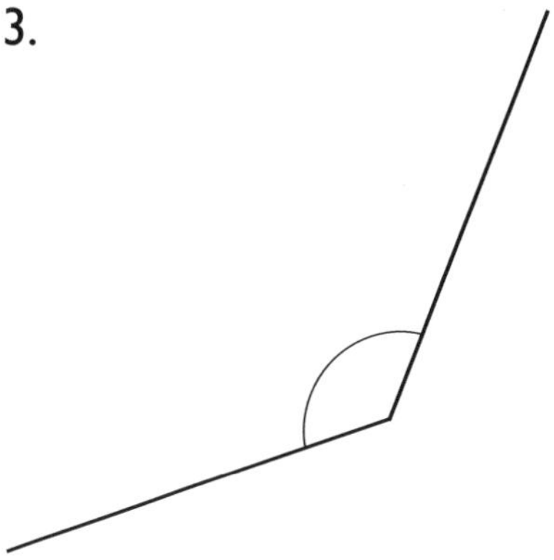

4.

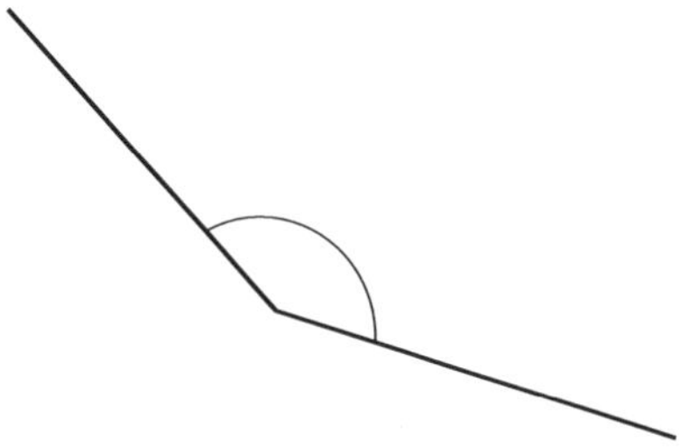

5.

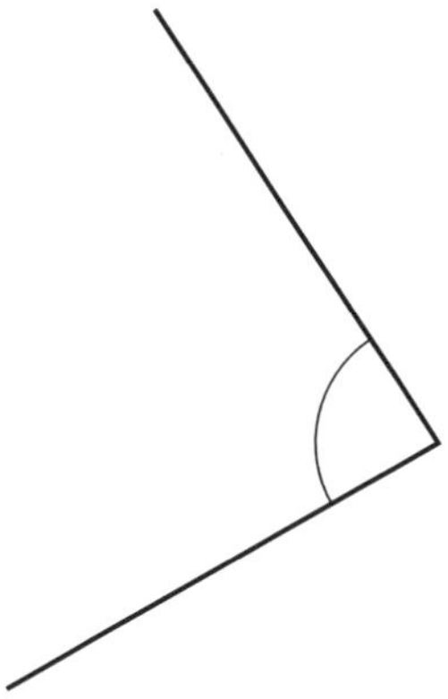

6.

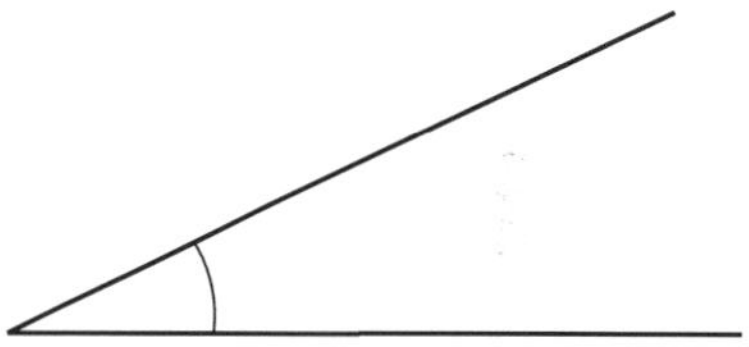

7.

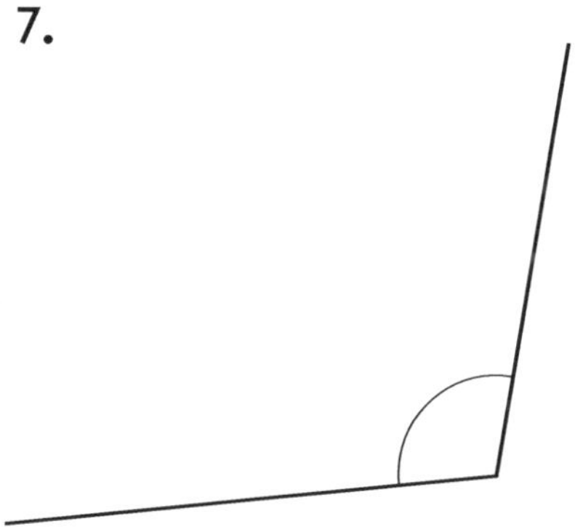

8.

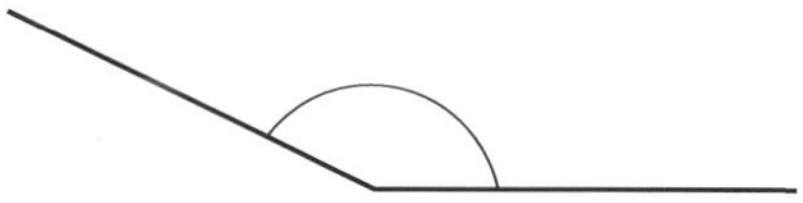

9.

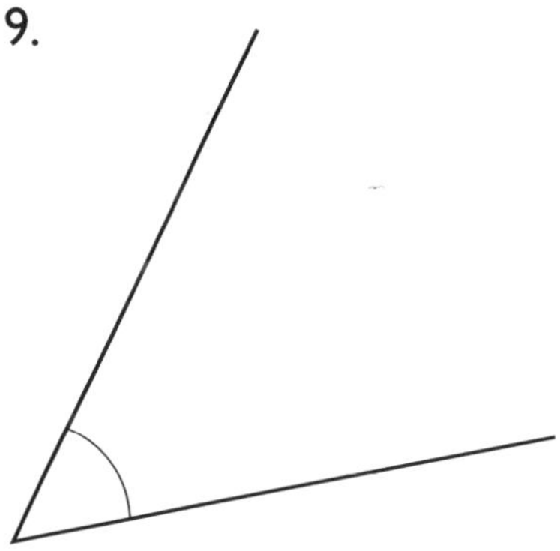

10.

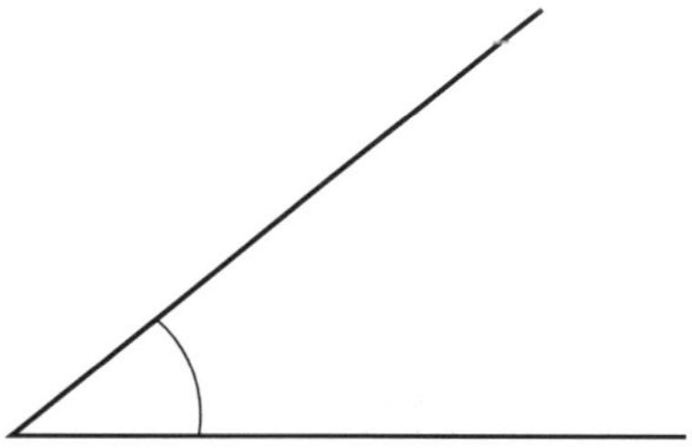

11.

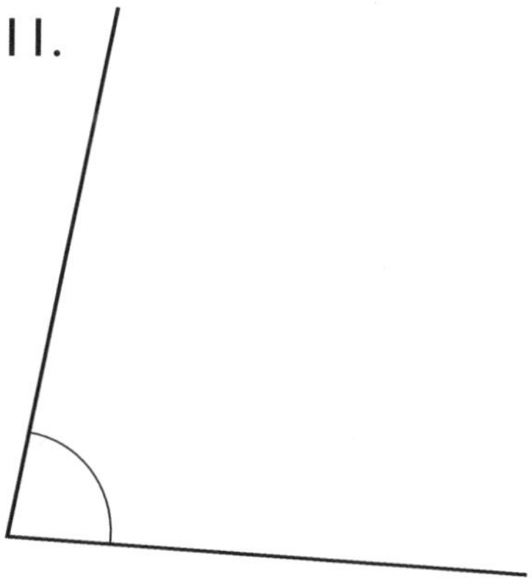

12.

Angles in a straight line cards

1.

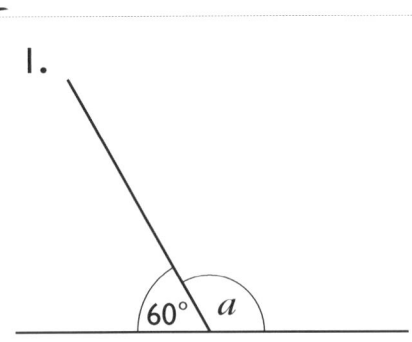

60° *a*

2.

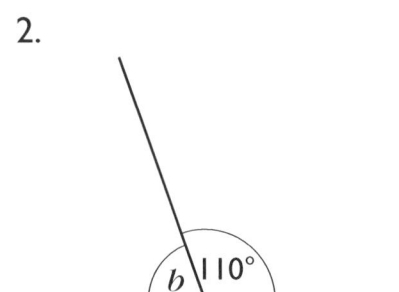

b 110°

3.

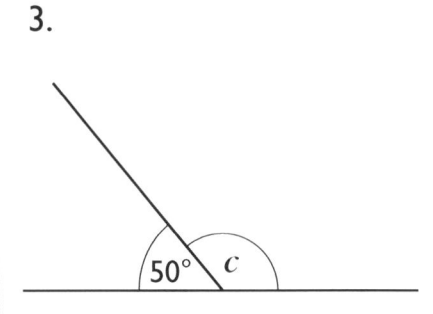

50° *c*

4.

35° *d*

5.

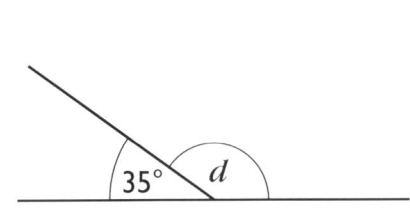

e 125°

6.

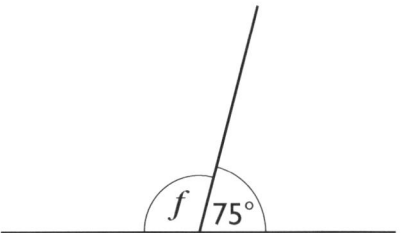

f 75°

7.

45° *g* 60°

8.

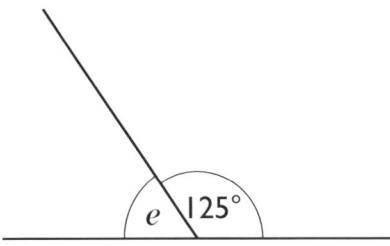

25° 90° *h*

9.

35°

i 85°

10.

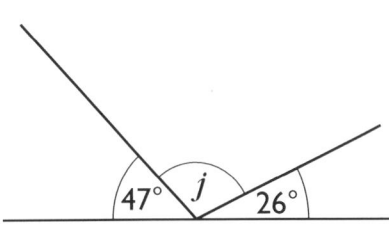

47° *j* 26°

11.

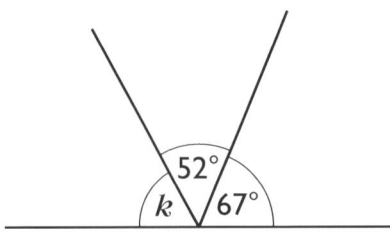

52°

k 67°

12.

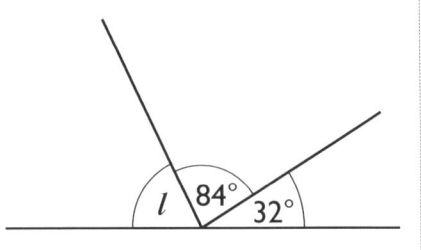

84°

l 32°

Converting measures cards

5·2 kg	8·7 l	3 km 900 m
460 cm	720 cm	50 mm
16·3 kg	3·6 l	18·5 km
7300 g	6800 ml	3 kg 400 g
47·2 cm	7700 m	6 cm 10 mm
34·1 km	18 300 ml	14·7 m
5800 g	2 l 400 ml	6 m 70 cm
82·9 m	58·3 cm	230 ml

Collins New Primary Maths

Scales

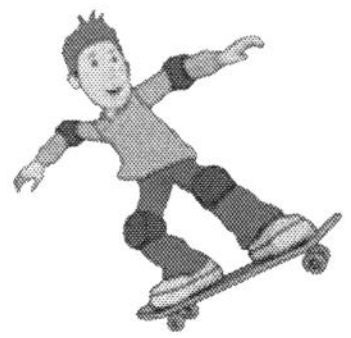

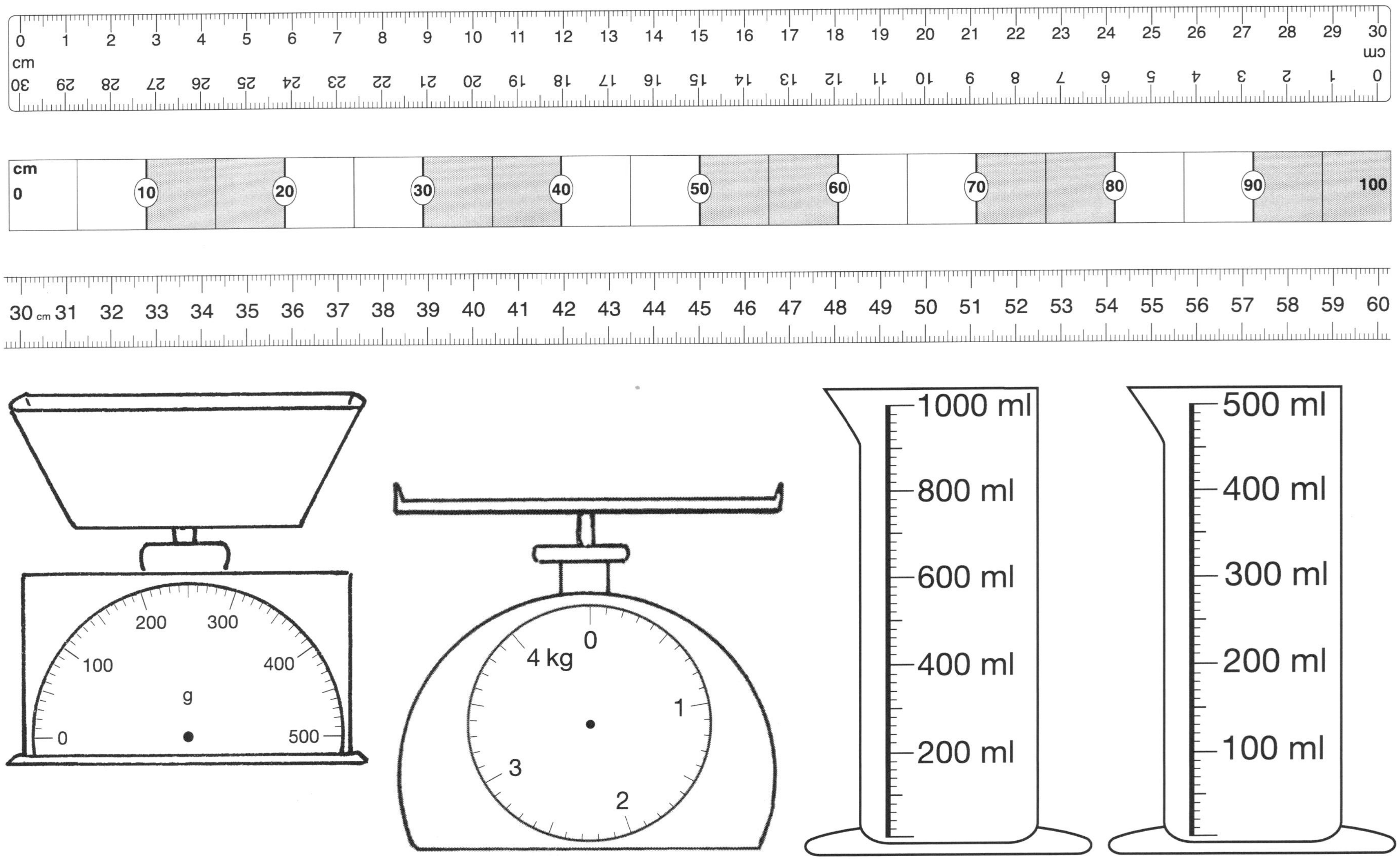

0 1 2 3 4 5 6 7 8 9 10 11 12 13 14 15 16 17 18 19 20 21 22 23 24 25 26 27 28 29 30
cm
cm
0
10 20 30 40 50 60 70 80 90
100
30 cm 31 32 33 34 35 36 37 38 39 40 41 42 43 44 45 46 47 48 49 50 51 52 53 54 55 56 57 58 59 60
0 100 200 300 400 500
g
0 1 2 3
4 kg
1000 ml
800 ml
600 ml
400 ml
200 ml
500 ml
400 ml
300 ml
200 ml
100 ml

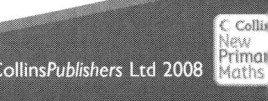

Polygon cards

1.

2.

3.

4.

5.

6.

7.

8.

9.

Collins
New
Primary
Maths

Rectangle cards

1.

2.

3.

4.

5.

6.

7.

8.

9.

Collins
New
Primary
Maths

Timetable and calendar

London to Penzance

London	0730	0905	1005	1205	1305	1405	1505	1605	1803	2350
Reading	0758	0932	1032	1232	1332	1432	1532	1632	1832	0035
Taunton	0952	1054	—	—	1450	1550	1700	1748	1949	0330
Exeter	1018	1125	1208	1408	1520	1618	1732	1816	2016	0404
Newton Abbot	1040	1146	—	—	1541	1644	1755	1839	2037	0425
Plymouth	1123	—	1306	1505	—	1727	1838	1920	2117	0508
Liskeard	1146	1301	1330	1531	1732	1751	1903	1948	2143	0631
Bodmin	1158	1312	1343	1544	1745	1803	1916	2000	2155	0645
St. Austell	1217	1331	1401	1559	1802	1824	1934	2019	2211	0708
Truro	1234	1349	1420	1618	1819	1841	1953	2036	2228	0727
Redruth	1247	1401	1432	1630	1830	1853	2005	2049	2241	0741
St. Erth	1305	1418	1450	1648	1847	1914	2023	2107	2255	0802
Penzance	1323	1429	1505	1705	1902	1932	2040	2125	2306	0828

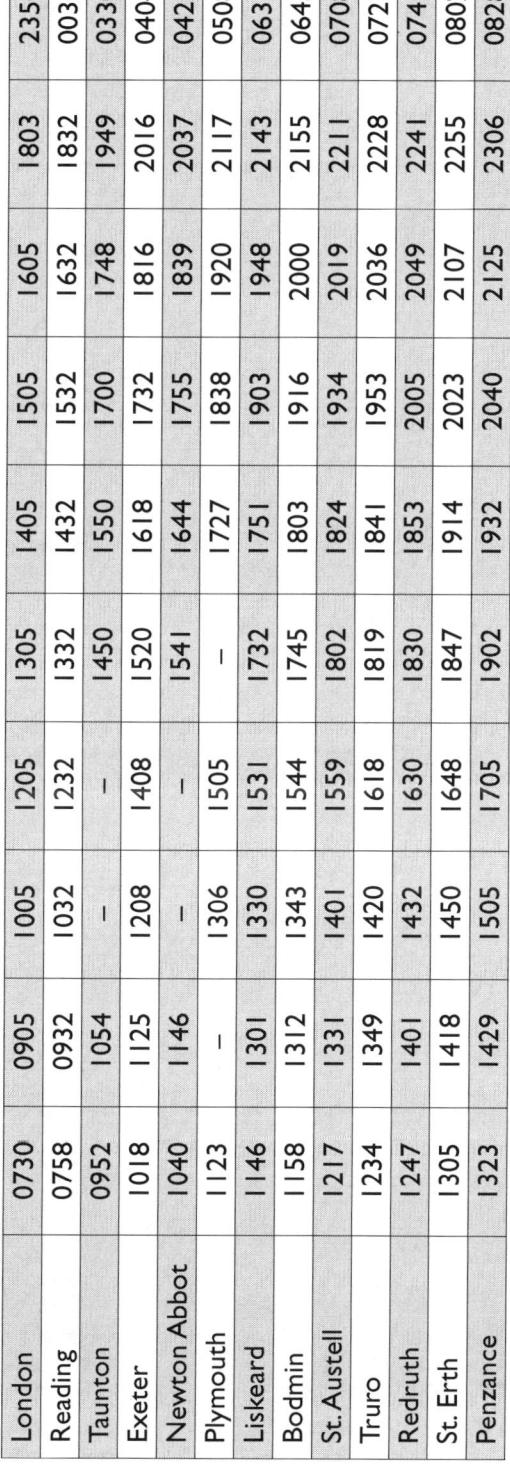

Calendar

JANUARY
```
S  M  T  W  T  F  S
            1  2
3  4  5  6  7  8  9
10 11 12 13 14 15 16
17 18 19 20 21 22 23
24 25 26 27 28 29 30
31
```

FEBRUARY
```
S  M  T  W  T  F  S
   1  2  3  4  5  6
7  8  9  10 11 12 13
14 15 16 17 18 19 20
21 22 23 24 25 26 27
28
```

MARCH
```
S  M  T  W  T  F  S
   1  2  3  4  5  6
7  8  9  10 11 12 13
14 15 16 17 18 19 20
21 22 23 24 25 26 27
28 29 30 31
```

APRIL
```
S  M  T  W  T  F  S
            1  2  3
4  5  6  7  8  9  10
11 12 13 14 15 16 17
18 19 20 21 22 23 24
25 26 27 28 29 30
```

MAY
```
S  M  T  W  T  F  S
               1
2  3  4  5  6  7  8
9  10 11 12 13 14 15
16 17 18 19 20 21 22
23 24 25 26 27 28 29
30 31
```

JUNE
```
S  M  T  W  T  F  S
      1  2  3  4  5
6  7  8  9  10 11 12
13 14 15 16 17 18 19
20 21 22 23 24 25 26
27 28 29 30
```

JULY
```
S  M  T  W  T  F  S
            1  2  3
4  5  6  7  8  9  10
11 12 13 14 15 16 17
18 19 20 21 22 23 24
25 26 27 28 29 30 31
```

AUGUST
```
S  M  T  W  T  F  S
1  2  3  4  5  6  7
8  9  10 11 12 13 14
15 16 17 18 19 20 21
22 23 24 25 26 27 28
29 30 31
```

SEPTEMBER
```
S  M  T  W  T  F  S
         1  2  3  4
5  6  7  8  9  10 11
12 13 14 15 16 17 18
19 20 21 22 23 24 25
26 27 28 29 30
```

OCTOBER
```
S  M  T  W  T  F  S
               1  2
3  4  5  6  7  8  9
10 11 12 13 14 15 16
17 18 19 20 21 22 23
24 25 26 27 28 29 30
31
```

NOVEMBER
```
S  M  T  W  T  F  S
   1  2  3  4  5  6
7  8  9  10 11 12 13
14 15 16 17 18 19 20
21 22 23 24 25 26 27
28 29 30
```

DECEMBER
```
S  M  T  W  T  F  S
         1  2  3  4
5  6  7  8  9  10 11
12 13 14 15 16 17 18
19 20 21 22 23 24 25
26 27 28 29 30 31
```

Collins New Primary Maths

Probability

likely

unlikely

certain

even chance

impossible

Collins
New
Primary
Maths

Collecting, selecting and organising data

Answer the following question by collecting, selecting and organising relevant data.

Things to think about

● How are you going to collect your data?
Perhaps using a tally chart or frequency table.

● How are you going to present your data?
Perhaps in a table or graph.

My initial ideas

How I am going to collect the data

How I am going to organise the data

Now do it!

What I found out

What else I could find out about this topic

Collins
New
Primary
Maths

Mode

3, 6, 5, 7, 3

10, 6, 3, 8, 6, 9, 2

6, 8, 5, 3, 5, 4, 2, 9

3, 7, 5, 8, 7, 4, 3, 2, 10

Die number	Tally	Frequency
1		
2		
3		
4		
5		
6		

Die number	Tally	Frequency
1		
2		
3		
4		
5		
6		
7		
8		
9		
10		
11		
12		

Die number	Tally	Frequency
1		
2		
3		
4		
5		
6		
7		
8		
9		
10		

Die number	Tally	Frequency
1		
2		
3		
4		
5		
6		
7		
8		
9		
10		
11		
12		
13		
14		
15		
16		
17		
18		
19		
20		